COMEBACK

COMEBACK

AMERICA'S NEW ECONOMIC BOOM

CHARLES R. MORRIS

PUBLICAFFAIRS

New York

Published in the United States by PublicAffairs™, a Member of the
Perseus Books Group
All rights reserved.
Printed in the United States of America.

PublicAffairs books are available at special discounts for bulk
purchases in the U.S. by corporations, institutions, and other
organizations. For more information, please contact the Special
Markets Department at the Perseus Books Group, 2300 Chestnut
Street, Suite 200, Philadelphia, PA 19103, call (800) 810-4145, ext.
5000, or e-mail special.markets@perseusbooks.com.

The Library of Congress Control Number: 2013936151
ISBN 978–1-61039–336–2 (PB orig.)
ISBN 978–1-61039–227–3 (EB)
First Edition
10 9 8 7 6 5 4 3 2 1

Contents

Introduction

I wrote a book called *The Coming Global Boom,* released in 1990, that pointed to the very bright prospects for the new decade—higher productivity, falling interest rates, an expanding finance and trading sector, and a recovery of leadership in computer technology. This was a very gloomy period, much like now. Good economists like Stephen Marris and C. Fred Bergsten at the International Economic Institute in Washington worried that the government's persistent deficits were about to ignite a worldwide recession. The economist/novelist Paul Erdman had a string of best sellers with names like *The Panic of '89.*

As a corporate banker and a business consultant, I was in and around a lot of different companies. I began to notice that most senior managers were in their forties, and that they were sitting atop hordes of bright, ambitious thirty-year-olds with almost no hope of ever moving up the executive ranks. I personally knew at least a dozen of them who left their companies to start a business—any business.

I started to run numbers on demographics and found some insightful academic work, especially by Richard Easterlin. Ed Yardeni was the rare economist who was thinking the same way. The thirty-year-olds of the 1980s were the same baby boomers who had depressed labor markets in the 1970s, but in the 1990s were entering the time of life when people *settle down*, get *very productive*, start *saving*. They had small families, and the generation of senior citizens was small as well. Play those shifts through the financial and housing and consumer goods markets and the new decade took on a real glow. The icing on the cake was that the United States had a near-monopoly position in the microprocessor technology that triggered the desktop and business server revolution.

Economists sometimes over-focus on things with clean numbers. To explain the burst of business formation in the 1980s most economists, especially conservative ones, point to a 1978 capital gains tax reduction, not a logjam of frustrated young talent trapped in corporate dead ends.

Today, the X-factor isn't demographics, it's the energy boom. Unless something goes horribly wrong, energy is a game changer, much as the demographic shift was in the 1990s. But this one may well last a lot longer. Here is the plan of the book.

Part 1 is entirely devoted to energy. I am not a starry-eyed enthusiast. I saw Josh Fox's *Gasland* only after I had learned a fair amount about the industry, and it struck me as mostly right. But I think the problems are fixable, and I will explain why. That discussion takes place in Chapter 3.

The Prologue to Part 1 and the first chapter attempt to give the reader a decent understanding of the industry, the geology, and the splendid technology that makes it possible.

Chapter 2 is a straightforward discussion of the job impli-
cations. I draw from a wide range of sources, but they mostly
agree that the total impact will be very substantial. I wind up
with a polemic against the rush to spend tens of billions to
send much of our bounty abroad. East Asia is desperate for
American gas, and they could easily take half or more of our
supply at very high prices. That would be great for the global
oil companies, but American prices would follow up the ladder and choke off the opportunity for a manufacturing boom.
It doesn't mean that we can never export, but the government
has been flooded with permit requests for exporting in volumes that could cause lasting price dislocations.

Chapter 3 is a fairly detailed dive into environmental and
related issues. The bigger companies seem to understand the
necessity for détente, and are spending a lot of effort to clean
up their acts. I hope they really mean it, because there is a lot
at stake.

In Part 2 I look at some important issues in the rest of
the economy, in the context of the political cycles that seem
to operate in the United States. Conservative and liberal as-
cendancies alternate in twenty-five- to thirty-year swings. I
voted for Ronald Reagan in 1980 because I was sure that the
liberal cycle that got underway in the late 1950s had run out
of steam and ideas, and that it was time to re-empower the
private sector. Now, I think we've reached a point where it is
the public sector that is impoverished, while the excesses of
the private sector have been paraded in an astonishing succes-
sion of scandals and crashes.

Chapter 4 is on infrastructure. Suffice it to say that it's
crucial, and a lot of it is falling down. American economic

recoveries are usually tied to infrastructure investment. Thanks to the advent of large amounts of cheap energy, we'll be able to afford it; the energy and manufacturing boom will require it; and it will keep the jobs machine rolling.

Chapter 5 is about health care, the industry everyone loves to hate. Why talk about health care in a book that describes an economic boom? Well, despite the waste and abuse in health care—though it could hardly be worse than in the global financial and oil industries—it is a vital economic lever that accounts for more middle-class jobs than any other industry I can find. It is also a big driver of the American leadership in semiconductors and biotechnology, and its success in treating the most serious diseases is far better than most people probably realize. But there is still a lot wrong with it, and I hope to shed some light on what and why.

The final chapter summarizes the scale of the opportunities, and offers a list of ways we could screw it up. The momentum is mostly going in the right direction, but could still be easily derailed. The next few years will pose critical tests of policy making and industry performance that could determine whether we have a broadly based economic revival or just make a few fortunes by feeding the Asian economic juggernaut. I've concluded with some thoughts on debt, growth, and taxes, because they all look different during an economic boom. That is the case for the comeback that is right in front of us.

Part I

THE MANUFACTURING/
ENERGY NEXUS

Shale Country

America's rise to global economic preeminence was built first and foremost on its splendid endowment of natural resources. From the 1870s, the great Midwestern "factory farms"—where fields could be plowed in mile-long straight lines—dominated world grain output. The Lake Superior iron deposits may have been the world's richest, and were ideal for steelmaking. Add the California gold bonanza and the great anthracite seams of Pennsylvania.

But oil was the quintessential American industry. Unlike ore and coal, oil wasn't wrested out of the earth in pickaxe-size bites. When you found an oil deposit and drilled into it, it gushed. And as John D. Rockefeller was the first to appreciate, with most transport by pipeline, oil scaled like no other industry. For a few decades in the nineteenth century, Standard Oil had nearly the whole world market to itself.

After Standard Oil was broken up, the industry's center of gravity shifted to the mid-continent region stretching from the Texas Gulf Coast through almost the whole of Oklahoma and into eastern Kansas. Much of this is rolling, open-sky country, and in both Oklahoma and Texas, the industry existed cheek by jowl with cattle ranching. They share a swaggering, high risk, brush-off-the-losses-and-start-over approach to the world and, especially in oil, the greatest gushers, along with sudden fabulous wealth, often came only after years of dry holes.

Oklahoma City is one of the epicenters of the shale-based gas and oil boom spreading through the geographic midlands of the United States. Some 400 million years ago, much of the region between the western slopes of the Appalachians and the eastern slopes of the Rockies was covered by inland seas. In geologic time, sedimentary mud was compressed into thin layers of relatively impermeable shale rock, trapping decaying organic matter, which was gradually transformed into the multiple hydrocarbon compounds that humans burn for artificial light and heat. Eons ago, we began to exploit the oil in surface puddles that seeped from the ground; just over a century and a half ago, we discovered that we could tap into deep-underground pools of oil and gas that had leaked from the shale formations into natural reservoirs. And now, within just the last couple of decades, we have learned how to wrest hydrocarbons directly from the shale, which could multiply America's recoverable hydrocarbons by a factor of five, wipe out our energy trade deficits, and give a big boost to our industrial competitiveness.

Devon Energy is one the bigger independent players in the shale gas and oil industry, and by market value, the biggest of the Oklahoma independents. Compared to the global oil majors like ExxonMobil and Shell, Devon is a midsize company. By most real-world measures, it's a giant, generating $9–10 billion in annual revenues, with free cash flow in the $4–5 billion range, and only modest debt. The skyline of Oklahoma City is dominated for miles by the new Devon headquarters, a soaring 52-story glass tower that opened in 2012. All leaders of red-state companies are fiercely resistant to federal regulation, and the board chairman of Devon, Larry Nichols—he and his father founded the company in 1971— is no exception. But he takes environmental issues seriously, and the company has won Environmental Protection Agency (EPA) and Bureau of Land Management awards for reducing emissions and other adverse environmental impacts.

On a bright morning in January 2013, I was in the back seat of an SUV driving west from Oklahoma City to visit some Devon wells at different stages of the drilling and recovery process. The driver was Tim Hanley, a media relations staffer with thirty years of local newspaper and public relations experience; Robert Brodbeck, a senior petroleum engineer was with us as a technical resource. He is in his fifties, pleasant, strongly built, with a ruddy, weathered face and a Hemingway beard. He answered questions carefully, completely, with a military-style, nothing-but-the-facts, taciturnity.

Such visits, I had discovered, are very hard to arrange— doubtless because of the controversies over shale gas and oil, and the use of hydraulic "fracking"—rock-fracturing techniques that are used to loosen the shale-bound products. I was on this one through the good offices of Robert Bryce, also on the trip. He's a native Tulsan and a long-time Texas resident, and a veteran reporter on the energy and oil business. He's also a fellow PublicAffairs author, and drew on his contacts to shoehorn me into this trip, then decided to come along himself to catch up on the newest rig technology. The gold star for making the trip happen, however, goes to Sarah Terry-Cobo, a smart, thirtyish reporter for the main Oklahoma business newspaper, the *Journal-Record.* She wangled the trip permission during a happenstance conference conversation with Nichols, so Robert and I were riding her coattails. Sarah is a self-described Okie who did her graduate work at Berkeley, worked as an investigative reporter in California, then came back to Oklahoma to take up the health and energy beats at her paper. We brought our own boots, and the company furnished hardhats, safety glasses, and fire-resistant smocks. Brodbeck looked like an exception to the fire-resistant rule, wearing a pressed flannel shirt and jeans; when I asked him, he showed me the labels. If you're around rigs all the time, you buy only fire-resistant work clothes.

Oklahoma City is on the eastern edge of the state's wheat belt, one of the most productive in the country. We were driving through rolling grassland, and crossed the Chisholm Trail just west of the city. There were constant markers of the importance of the oil and gas business—chain-linked acres of stacked pipe; parking yards for giant sand trucks; a Devon gas compression and purification plant; rigs sticking up here and there in the grasslands. Near the end of the drive, we traveled a short distance on the old Route 66, the road that John Steinbeck's Okies traveled fleeing the Dust Bowl tragedy of the 1930s. The state has preserved an original segment, just two strips of concrete, with an irregular center line, no painted markings, and no shoulders. The shale underlying this part of the state is called the Cana-Woodford, a particularly favored location since it is rich in natural gas liquids (NGLs) that can be converted into many of the same products as those from crude oil.

After about forty-five minutes, we shifted to hard-packed clay back roads, laid out in perfect squares, and finally we turned onto an asphalt road with a row of ten identical rigs each about a quarter-mile apart. Almost all the independent Exploration and Production (E&P) companies contract out drilling and well completion to oil field service companies, like Halliburton and Schlumberger, which are among the biggest worldwide. The driller specialist Devon was using on these rigs, Helmerich & Payne, is another Oklahoma company with international reach, and like Devon, illustrates what a fast ride these firms have enjoyed. In 2000, H&P had seventy-six active rigs roughly split between the United States and overseas. Revenues were $351 million with an operating profit of $46 million. In 2012, with 336 rigs, 240 of them active in the United States, revenues were $3.2 billion, with $910 million in operating profit. The company has almost no debt and threw off more than $1 billion in free cash, all of which went to capital expenditures.[1]

Shale drilling poses unique challenges. The shales are often very deep underground and, although they extend for hundreds of miles horizontally, they are typically only hundreds of feet thick. In contrast to conventional oil and gas wells, the hydrocarbons in shale are distributed thinly, so in order to be cost-effective the well has to access a very large area of the shale. All ten of the wells on this site will drill down about 12,500 feet then make a ninety-degree turn and extend horizontally through the shale for another mile. The horizontal portion of the pipe is perforated; once the shale is fractured, hydrocarbons will flow into the pipe all along its length, driven by the pressures from the gas in the shale. A rig site, or a pad, can usually accommodate eight to ten wells, but the ten pads at this site will each drill only two or three, due to nuances of the geology.

H&P is the highest-cost driller on a day-rate basis, but still the most cost-effective because of the speed and accuracy of its drilling. It designs and manufactures its own brand of rig, the "FlexRig." The rig is powered by electricity generated by three large diesel engines. Electrical power allows both higher speeds and precise adjustments of drilling speeds and pressures as the drill progresses through the geologic strata. The FlexRigs—all 136 feet of them—are moved with dozers and cranes from wellhead to wellhead on the same pad, with the platform and the rest of the equipment repositioned around them. H&P has a new line of FlexRigs that are moved by hydraulic pistons on prepositioned tracks to speed up the operation even further. The pad we were visiting had three wellheads quite close together, but as the wells are sunk, the drills are maneuvered so the collection (horizontal) pipes are arrayed in parallel lengths several hundred yards apart.

Brodbeck gave us a brief tutorial on directional drilling. The drill bits—they're diamond studded and cost about $20,000 apiece—are clamped into pipe-length holders, and turned by a rotor apparatus in the holder that is driven by the force of the drilling mud. To change the direction of the hole, the bit holders are changed for ones that are curved at the end by just a degree or two. It takes about a quarter mile of depth to execute the turn from the vertical to the horizontal. The steel casing pipes are sufficiently flexible that they bend to fit the shape of the hole as they move through it. Only a cable containing electrical wires connects the drilling apparatus to the rig.

There were six H&P employees on site. The driller runs the rig and supervises the derrick man, the motor man, and two roughnecks. They all work twelve-hour shifts for seven days, seven days on and then seven days off. The sixth man, the tool pusher, has the same seven-days-on, seven-days-off tours, but stays on location throughout his days on. Devon is represented by a drilling superintendent and engineer team for each three or four rigs, plus a drilling manager per rig. Since the rigs operate 24/7, four men are needed for each position. The drilling manager at this rig, Tim Taylor, is a great cheerful bear of a man, I'd guess in his mid-fifties. His career has taken him throughout the world, with significant spells at the Aramco sites in Saudi Arabia. He enthused over the new drilling rigs—drilling times have been cut in half compared to just a few years ago. Tool and pipe handling is now very mechanized. The pipes come in thirty-foot lengths. They were once craned up to the drilling platform and wrestled into racks by the roughnecks but are now automatically loaded onto a conveyor belt, stood up in the rig, and connected into ninety-foot "threefers," all by machine, reducing the number of pipe changes.

Oil and gas field workers are typically big men, heavy around the neck and shoulders. Although mechanization has eliminated a lot of the heaviest work, it still requires a lot of fast, strenuous, coordinated action. Pipe changing is one of the most routine tasks. As the drill pushes to the limit of its pipe, two roughnecks position themselves on the platform and, using giant ratchet wrenches hanging on chains, disconnect the well pipe from the drilling mechanism, swing in a new length of pipe, and lock it into place, all in about thirty seconds, all the time getting spattered by the cascade of drilling water and mud from the disconnected pipe. (A drain pan captures the fluids and they're recycled.) Twelve hours of that, plus all the routine rig maintenance tasks, adds up to a solid workday, with no space for shirkers.

The drilling cab housing is the nerve center of the operation. Much of the drill routine is preloaded into the computer, and an array of digital displays tracks the progress both visually and numerically. But since nothing ever goes completely as planned, the driller plays a crucial role, rather in the way an airline pilot earns his pay during high-stress moments. The driller for this rig, Artie White, a young man of about thirty, has been with H&P for eight and a half years. He worked every rig job for the first three years before being promoted to driller. When we were there, a tool had broken in the well—it happens—and Artie was carefully backing the drill out, using mud to clear any small metal filings from the break. Because the shifts are so long, I asked Tim about lagging attention during long periods of computer control, a well-known issue for airline pilots. He explained that there are arrays of alarms if a value varies outside of the programmed range, and the driller's displays, as well as the alarms, display throughout the company on its internal web. Tim pulled them up on his Smartphone for me. So a driller is never completely on his own. Later I asked Norm Naill, Tim's boss, what a driller got paid. He guessed between $28 and $32 an hour, with good benefits. Part of the 84-hour duty stretches also counted as overtime, so the gross annual pay would come out to between $77,000 to $88,000, with perhaps a 10 percent bonus on top of that. The roughnecks' annual pay was lower, of course, but not that far behind. So these were all young guys making a very good living on the ground floor of an industry that looked to keep expanding for a long time.

After a couple of hours at the rig, as we were getting ready to leave, I was standing with Tim, and he beamed, "Charlie, I can't believe I get paid to do this work. I'd be happy to do it for free. I've been doing it for a lot of years and I still love every minute of it!"

The next stop on our tour, about a half hour away, was a well in the "completion" stage—actually fracturing the rock and collecting the product. There were three wells on the site, in the form of three stubby, capped pipes sticking up out of the ground, festooned with connection points, valves, and gauges. (The trade calls them "Christmas trees.") Fracturing takes place in stages. The horizontal well pipes on these wells were about a mile long, and they were doing ten different fracturing events, each one for a 500-foot portion of the shale. The entire fracking process for each well was assumed to take thirty-six to forty-eight hours, on the usual 24/7 two-shift workday schedule. Halliburton was the completion contractor, and everyone on the site wore the company's bright red coveralls.

The well pad was in a flat, excavated area about fifty yards long, with a further excavated area behind it. There was an array of control trailers where we entered, then the rig and the three wells in a tight row, and to their left several closed sand trucks and chemical tanker trucks. Four six-inch pipes ran on the ground from the first wellhead through the length of the site, passing between an array of eighteen forty-eight-foot-long red Halliburton trucks arrayed tail-to-tail in two neat nine-truck rows. Each of them was a 2,000-horsepower diesel engine with a cab and wheels. Behind them was a mixing tank called the "blender," and behind that in the second excavated area, ten water tanks. The average Devon well on this site, with three-plus miles of well pipes per well, takes about 7.5 million gallons of water, including drilling and fracking, or roughly twenty-three "acre-feet" of water—picture a square one-foot-deep pool measuring about 1,000 feet on one side.

Bright blue lines of flexible irrigation tubing snaked through the grassland to the tanks, bringing water either purchased from local farmers or recycled from Devon's own treatment facility. During fracking, water is pumped into the blender where it is mixed with a variety of chemicals to reduce friction and keep the pipes clean, as well as sand of a specific grain size, to prevent fractures from reclosing. One of the big thirty-foot-long sand trailers had been stood on its front end and locked into a sand dispensing machine; it looked like a prehistoric reptile that had stumbled headfirst into a tar pit. The proportions and mixing speeds of the chemicals and the sand were all computer controlled.

When it's time for the actual fracking, the fluid, or "slurry," flows through the four pipes between the eighteen diesels, with the capacity to generate 12,000 psi inside the well (but 2,000 psi of "backpressure" is withheld as a safety margin).

Apart from the airport-scale noise of the eighteen trucks generating 36,000 diesel horsepower and the visible vibrations in the feed pipes, there's not much that is noteworthy about a normal frack. The liquid flow is in a closed system—from the tanks through the blender and the powerful gauntlet of diesels into the well, with the bounceback captured in the drainage tanks.

We were at the fracking site for about two hours, so we took potluck on what we saw, which happened to be the perforating of the stage, an intricate and fascinating technology of the kind that evolves over many years with microcontributions from hundreds of craftsmen. The perforating gun tubes were studded with stainless steel openings about an inch wide that fired shaped charges to punch holes through the steel and cement casing.

The team was working on the fourth stage of a frack when we arrived, and we stayed until the perforating guns were in position. Another well-paid thirty-something sat in front of an array of digital displays, much like those in the drilling cab, scrolling rapidly through length markers of the well pipe. As he landed on the proper location, he would say "firing," a red square would flash on the screen, and an adjoining screen would show a visual pressure wave in the shale. A consulting engineer representing Devon marked each firing and its numerical location. Not taking any chances, a roughneck outside the trailer held the tautly stretched control wire in both hands, confirming each time that he had felt the vibration.

The next work phase would involve extracting the used guns from the well, cleaning any debris, and setting the plugs to isolate the fracking site, all before the pumping started.

Because of the length of the pipe, Brodbeck estimated that it would take 2.5–3 hours to complete the fracking. But we were running out of time, and so we headed off for the last station, the first Devon fracking water recycling plant.

Central Oklahoma is typically well-watered, Brodbeck told us, pointing to the treeline of the North Canadian River visible in the distance. It is a distant tributary of the Arkansas River, is typically fast-flowing, and feeds a system of recreational lakes created by Oklahoma City. But the state, like much of the southwest United States, had been locked in a three-year drought, and concern was growing about falling water tables.

All shale drilling sites have holding tanks for used fracking water. Deep rocks are full of noxious elements, and fracking water's violent trip into and out of the rocks leaves it heavily contaminated. Devon is one of the first companies to create their own centralized collection and treatment facilities in order to reuse both fracking and produced water. Brodbeck estimated that in Cana-Woodford, they recover 40–50 percent of the water and can reuse 90 percent of it. Assuming 45 percent recovery, that would allow them to extend their water supply by about two-thirds. Recovery rates vary considerably by local geology and often even from well pad to well pad.

The operation would not be feasible without a dense concentration of wells, like Devon's in the Cana-Woodford. The center of the operation is a twenty-one-million-gallon plastic-lined holding field—a small lake—to hold the treated water. Water is trucked in constantly from the Devon wells in the area and treated through a standard filtering and sedimentation process. The clean water drains into the lake, and is tapped by irrigation pipes like those we had seen at the fracking site. The 10 percent of the water that is not reusable is a fairly noxious mixture of salts and metals. But the volumes are modest, and the company has dug a disposal site, where the waste is injected thousands of feet under the surface, that will serve, one hopes, as its permanent home.

Devon hasn't released the exact cost of the facility, but it certainly cost the millions the company claims. Since water is badly underpriced throughout America, the plant could not be cost-effective, but it is the right thing to do, and at some point should probably be required of all companies. And Devon deserves credit for staying in front of the curve.

CHAPTER 1

An Unconventional Revolution

F orty years ago there was a seismic break in the geography of world power. A dozen countries that produced much of the world's oil—almost all of them in the Middle East— declared an oil embargo against the West. They were retaliating for the humiliation of the 1973 Arab-Israeli war, but also announcing their independence from the American and European global oil companies that had long controlled their production and prices. Oil prices tripled (from about $3 to about $10 a barrel), and the United States was plunged into the nastiest recession of that time since the Great Depression. Runaway inflation in the United States and a collapsing dollar prompted another tripling of oil prices in 1979 and the steep recession of 1981–1982.

The economic collapse combined with the end of oil price controls in the United States caused a sharp drop in oil demand, and oil prices dropped steadily through the middle 1980s. Memories of the crisis faded when prosperity returned in the 1990s, and the faux American boom of the 2000s brought a return of the old profligacy, as soccer moms ferried their kids in vehicles modeled after cattle-country pickups and Gulf War troop carriers. More and more of US manufacturing migrated overseas, and the trade deficit, about a third of it energy-related, went off the cliff. In the run-up to the 2008 financial crash and Great Recession, crude prices scraped $100 a barrel. Nightmare visions of a 1970s replay seemed all too plausible.

Behind the scenes, however, awareness was slowly spreading that the United States was swimming in inexpensive energy. As the reality of the American energy position sunk in, engineers stopped work on giant multi-billion-dollar natural gas importing facilities in Louisiana and Texas, and began reconstructing them to *export* gas. In the fall of 2012, the International Energy Agency (IEA) forecast that by 2020 the United States would surpass both Saudi Arabia as the world's largest oil producer and Russia as the world's largest gas producer.[1]

The new American energy bonanza stems primarily from "unconventionals," land-based hydrocarbon deposits that geologists have long known about but which were considered inaccessible on both technical and economic grounds. The unconventional energy bonus has also been supplemented by the steady expansion of the industry's deep-sea drilling capabilities.

Unconventional Hydrocarbons

Coal, oil, and natural gas all derive from decayed plant and animal matter that has slow-cooked within the earth for hundreds of millions of years. As dead organic material accumulated on the muddy bottoms of swamps, lakes, and oceans, the mud sank and was gradually compressed into sedimentary shale rock, trapping the organic matter within its many layers and fault lines. Compression generates heat, which gradually transformed the organic matter into ordered chains of carbon and hydrogen atoms. The simplest chain, one carbon and four hydrogen atoms (CH_4) is methane, the premier heating gas, and one of the most abundant organic compounds on earth. The hydrocarbons that are used to manufacture gasoline are much heavier, with four to twelve carbon atoms and multiples of hydrogen atoms; those used to make diesel fuel are heavier still. The hydrocarbons favored for oil and gas production tend to be derived from the decay of phytoplankton in marine (saltwater) settings and of algae in freshwater lakes, while coal is thought to derive from swampy peat forests like those still found in some southeast Asian countries.

Shale is therefore called a source rock—it traps and transforms hydrocarbons. By itself shale is nearly impermeable, but over the vast stretches of geologic time, liquid or gaseous hydrocarbons seep from between the shale layers into the surrounding earth. Conventional oil reservoirs form when a hydrocarbon-rich source rock leaks into a highly permeable medium like sand formations—the "reservoir rock." If there is another layer of shale over the reservoir, it will act as "cap rock," sealing the seepage in the reservoir. Drillers harvest the oil or gas by sinking a pipe through the cap rock into the reservoir, and sucking it out as if with a straw. For traditional oil and gas exploration companies, the hard part is finding the reservoirs; although geologic surveying technology is improving at a rapid rate, most oil exploration efforts still end up with dry holes.

Shale, however, can serve as both a source rock and a reservoir rock. If the openings between the horizontal layers are wide enough it is always possible to leach out usable hydrocarbons. But the economics have been unattractive. Shale formations are often very deep in the earth, so drilling is expensive. Even in the best shale, hydrocarbons rarely make up more than 6 percent of the rock weight. And although shale formations often stretch over tens of thousands of square miles, they're typically only a few hundred feet thick. Drilling that far down to access such a paltry area of shale, with such a small payload of hydrocarbons, made no sense.

Shale is not the only source of unconventional gas and oil, although it is by a large margin the game changer for the United States in the near term. There are at least three others: coalbed methane, or natural gas locked up in the seams of coalbeds; oil sands—heavy, rock-like concentrates of hydrocarbons permeating sand formations; and oil shale—which is not shale and has nothing to do with oil-bearing shale formations. Coalbed methane and oil sands are being exploited now, while oil shale is still quite far from commercial exploitation. The recovery technologies are described at the end of the chapter. The rest of this chapter will focus on shale-based gas, natural gas liquids, and oil, where the main opportunities lie.

The Technology of Shale Drilling

The American shale revolution was born from a convergence of technologies. Two critical ones—directional drilling and well stimulation by hydraulic fracturing—were already well established, but needed to be adapted to shale. Both the industry and the government had excellent geologic maps of underground resources, and good information on the most likely shale opportunities. Other critical technologies were the software, hardware, and tools required to track and precisely direct drilling tools deep underground and to monitor wells and fractures.

The political road was eased by the fact that the first important shale discoveries were in lightly populated communities, and in America, unlike in most other countries, landowners almost always own the associated mineral rights. Once shale drilling became technically feasible, therefore, companies could quickly assemble the large areas of underground rights required for profitable shale production.

Credit for working out the final engineering solution usually goes to George P. Mitchell of Mitchell Energy Co., who invested about a decade and $6 million of his own money to create the first complete replicable technical protocol for successful shale exploitation. Mitchell's first successful horizontal well, in the Barnett shale, right outside Fort Worth, Texas, was completed in 1991. In 1998, after successfully solving the challenge of mass hydraulic fracturing with "slippery" polymers (to lower the friction in the well pipe), Mitchell achieved the first commercial production of shale gas. Mitchell retired in 2002, after selling his company to Devon Energy. The Barnett is now the most intensively developed shale gas and gas liquids site in the world.

Mitchell richly deserves his honors for making shale exploitation a practical business, but the conventional account usually leaves out the role played by the federal government in the development of the essential technologies. In the midst of the oil price crisis of the 1970s, President Gerald Ford launched the Eastern Gas Shales Project to develop shale gas extraction technologies through demonstration contracts. In parallel, the agency that is now the Department of Energy joined the industry in funding research and demonstrations by the industry-sponsored Gas Research Institute. The government's Morgantown Energy Research Center (now the National Energy Technology Laboratory) was already involved in a substantial coalbed methane program, which contributed to the same technologies that enabled shale exploitation. The essential three-dimensional seismic imaging technology (to track the progress of drilling) was developed at the Sandia National Laboratory, originally for use in coal mines, while researchers at Morgantown made important contributions to drill bit construction and directional drilling technology. Mitchell himself was a leader in pressing the case for government support, and his successful 1991 well was partially funded with federal research and development (R&D) money. In addition, in 1980 the Carter administration and the Congress passed a special production tax credit for unconventional gas (this expired in 2002 when the industry had achieved solvency).[2] Altogether, the joint federal-industry development program in unconventional energy was pursued in much the same spirit as that of the early semiconductor industry.

I will describe the key technologies below, while reserving issues of safety and environmental and related issues for Chapter 3.[3]

Drilling and Casing the Well

Directional drilling is essential because of the typically narrow vertical profile of shale formations. A shale drilling rig sinks a well vertically until it nears the target shale formation, then, like the rigs used for the Devon wells described in the Prologue, it gradually turns the drilling tool so it travels through the shale in a more or less horizontal plane, and even maneuvers the drill to follow the contour of the shale. Wells are routinely drilled to depths of two miles or more, and then extended horizontally through the shale for as much as another two miles. To compensate for the thin distribution of hydrocarbons, shale well pipes are perforated so they can absorb hydrocarbons along their entire horizontal length. And since the shale formations typically stretch in all directions, a single rig, or "pad," may sink up to ten wells to cover the accessible shale terrain. An additional bonus of working in shale is that the distribution of the hydrocarbons tends to be fairly uniform over broad areas, so dry holes are a rarity. The economics are more like those of a manufacturing operation rather than of conventional gas and oil drilling.

Wells are sunk in stages. In the first stage, a large diameter hole is sunk to a depth of up to 150 feet. A heavy drilling mud or other fluid flows in behind the drill to stabilize the walls of the hole. Pipe casing, usually in thirty-foot lengths, is inserted as the drilling proceeds, and as each casing stage has been placed, cement is forced down the casing under sufficient pressure to force it out the bottom and back up around the sides to fill the space between the hole walls and the exterior of the casing. That first segment is called the "conductor

casing," since it sets the vertical direction and serves to stabilize the subsequent casing insertions.

The next stage is the "surface casing," a narrower pipe that
fits within the conductor casing. The surface casing is always
sunk to a depth below any intervening drinking water aquifer.
Depending on the depth of the well, several further "intermediate" and "production" casings are inserted and similarly
stabilized with cement. The final production casing is usually
only four to six inches in diameter in order to keep the volumes of required mud and cement within manageable proportions. (See graphic on page 27.)

Fracturing

Hydraulic fracturing is necessary to stimulate the release of
hydrocarbons from the shale, by releasing a high-pressure
burst of fluid—or sometimes compressed air or a hot gas—into
the shale. The pipe must first be perforated, using perforating
gun tubes. The tubes come in a variety of configurations: the
ones used at the Devon well we visited were studded with
stainless-steel covered openings about an inch in diameter,
forty-eight in each eight-foot tube-length, one tube to a
stage—and, with ten stages, 480 collection perforations in all.
The powerful explosives are packed into bazooka-shell-like
shaped charges to concentrate on a small point. The leading
tip impacts at 25,000 to 30,000 feet per second generating
ten to fifteen million pounds per square inch. Those pressures
"overcome" casing and cement strength—basically, they melt
it—and open a clean hole into the shale, penetrating a foot
or so beyond the tube. The tube construction also embodies
sequenced magnetic and nonmagnetic materials that can be

Typical Fracturing Operation

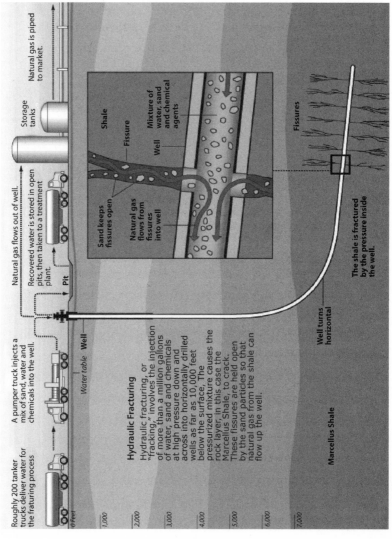

Roughly 200 tanker trucks deliver water for the fracturing process

A pumper truck injects a mix of sand, water and chemicals into the well.

Natural gas flows out of well.

Recovered water is stored in open pits, then taken to a treatment plant.

Natural gas is piped to market.

Storage tanks

Pit

Water table   Well

Hydraulic Fracturing

Hydraulic fracturing, or "fracking," involves the injection of more than a million gallons of water, sand and chemicals at high pressure down and across into horizontally drilled wells as far as 10,000 feet below the surface, The pressurized mixture causes the rock layer, in this case the Marcellus Shale, to crack. These fissures are held open by the sand particles so that natural gas from the shale can flow up the well.

Well turns horizontal

Marcellus Shale

The shale is fractured by the pressure inside the well.

Sand keeps fissures open

Natural gas flows from fissures into well

Shale

Fissure

Well

Mixture of water, sand and chemical agents

Fissures

0 Feet
1,000
2,000
3,000
4,000
5,000
6,000
7,000

Source: EIA

interpreted by external sensors, allowing the controllers to place the tubes with accuracy of within an inch or so.

The typical fracturing fluid consists almost entirely of freshwater, plus a small volume of sand called a "proppant," sifted to a precise size to prop open the newly created fracture channels, and a cocktail of chemicals. The mix of chemicals will vary by site and geology, but it commonly includes friction-reducing polymers and acids to clear perforations and to abrade and widen the fracture openings. The chemicals usually make up only about a half of one percent of the fracking fluids, but since a single well usually requires between three and six million gallons of fluid, some 15,000 to 30,000 gallons of chemicals will be applied to each new well. Assuming ten wells are sunk from a drilling pad, each pad area will absorb between 150,000 and 250,000 gallons of chemicals.

Given the miles of pipe to be traversed and the forces required to fracture shale rock, the fluid is injected under great pressure by trailer-truck-size diesel engines—the Devon site had eighteen—that generate up to 15,000 pounds of pressure per square inch (psi). Newly available software and hardware can generate real-time pictures from microseismic data showing the path of the fluid into the rock fractures. Chemical markers help trace the distribution of the proppant and identify areas of blocked holes in the perforated pipe. Once the fractures open, the pressures within the shale will immediately start the flow of hydrocarbons into the well pipe through the perforations.

The violence of the fracking makes it essential that well and casing pressures be carefully monitored during and after

Conventional and Unconventional Oil and Gas Recovery

Source: EIA

the operation to identify cement breaks or other leaks. Any fugitive, or uncontained, pressure outside of the well pipe is cause for heightened alert, and sustained pressures probably indicate liberated gases migrating up the perimeter of the well. The first rush of gas up the well pipe will bring considerable flowback of the fracking fluid, which will have been contaminated with heavy metals and other potentially toxic substances from the rock fractures, including naturally occurring radioactive materials or NORMs. (The organic matter in shale sometimes binds uranium and its derivatives.)[4] The fracking fluid is then followed by "produced" fluid, water that has been trapped in the rocks for eons, which is frequently more noxious than the fracking fluid. Good well management requires the sequestration and safe disposal or recycling of both the fluids and their associated hydrocarbons.

The Geology and Geography of Shale

When organic matter is trapped under sedimentary rock and gradually pushed deeper into the earth, microbial action and increasing heat expel oxygen, microbial methane, and other contaminants. If compression continues and temperatures rise to about 120°F, the organic material congeals into an insoluble substance called kerogen, the source of all hydrocarbon fuels. As temperatures continue to mount, all noncarbon material, including the hydrogen, is gradually stripped away; eventually as temperatures rise over 400°F, there is nothing left but an inert mass of nonreactive carbon. The different stages of the kerogen's transformation generate various hydrocarbons of commercial interest, with the earlier stages typically being the heavier and binding the most contaminants. Later stage productions include the "wet" gases that are sometimes liquid at atmospheric pressures, and generally have fewer contaminants. Methane is the smallest stable hydrocarbon molecule and the last usable kerogen product and is often found in a nearly pure state.[5]

Geologists have been examining the US shales at least since early in the twentieth century. They were long understood to be a source of valuable commodities, including uranium, as well as oil and gas. Analysis of the shales generally proceeds by sinking boreholes, an expensive process given the depth of many shales, and acquiring samples for chemical and other testing. The presence of recoverable hydrocarbons is deduced from a range of indicators. The shale's thickness and its thermal maturity—a heat index that suggests its kerogen's stage of transformation—can both be measured directly. Other important data include the shale's total organic content, usually

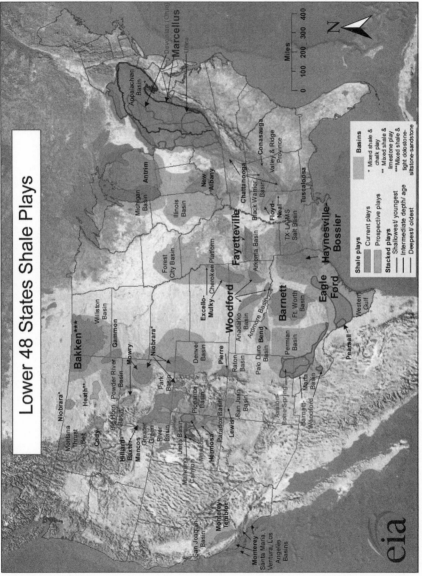

Lower 48 States Shale Plays

Shale plays
- Current plays
- Prospective plays

Stacked plays
- Shallowest/ youngest
- Intermediate depth/ age
- Deepest/ oldest

Basins
- Mixed shale & chalk play
- ** Mixed shale & limestone play
- ***Mixed shale & light dolostone- siltstone-sandstone

Source: ELA

expressed as a percentage of weight (4–6 percent is a good number); the amount of adsorbed gas; the volume of free gas within pores and fractures; and the source rock's permeability. Since those data are hard to measure directly, they are usually derived from sophisticated mathematical models, using radiometric data, patterns of magnetism, satellite thermal images, three-dimensional seismic analysis, as well as empirical data from similar developed shales. Gamma radiation, for example, can be a good indication of the quantity and nature of the kerogen.[6]

The map on the previous page shows the major shale "plays"—shale areas under active exploitation in the continental United States. Shales predominate throughout the middle sections of the country, covering much of the area between the western slopes of the Appalachians and the eastern slopes of the Rockies. The geologic record suggests that they are the sedimentary relics of large inland seas that inundated the region some 450–400 million years ago. They can be very large—the Marcellus shale, which stretches from western New York State to Ohio and southward to Virginia, covers an area of more than 100,000 square miles. The depth and thickness of shales also vary quite dramatically. The depth of the Marcellus undulates from 4,000 to 8,500 feet below the surface, and its thickness varies from 50 to 200 feet. With total recoverable reserves of 141 trillion cubic feet (Tcf), the Marcellus is one of the world's largest reservoirs of "dry gas"— almost pure methane that, with only minimal processing, can be piped directly to electrical utilities and to heavy industrial establishments to power their generators and machinery.[7]

But the Marcellus is just one part of a massive shale system, the Appalachian Basin that extends from the Lake Ontario region down to Kentucky. The Utica shale lies under the Marcellus, but extends further in every direction and may have twice the area. It is also thicker (ranging from 100 to 500 feet) and much deeper (from 2,000 to 12,000 feet). Over the last couple of years, the Utica has attracted a great deal of drilling interest because it produces heavier products than the Marcellus—natural gas liquids (NGLs) like propane, butane, and "natural gasoline"—that currently command a higher market price than dry gas. A 2012 survey by the United States Geological Survey (USGS), known for its conservative analyses, concluded that the Utica had recoverable reserves of thirty-seven Tcf of natural gas, or the rough energy equivalent of more than six billion barrels of oil,* plus an additional 1.2 billion barrels of recoverable oil and NGLs.[8] A third large formation, the Devonian, extends from the southwestern edges of the Marcellus further south and west into Kentucky, Virginia, and West Virginia, in several varying formations with an area of about 75,000 square miles, with substantial reserves of both gas and NGLs.

The so-called mid-continent region shale formations, the Fayetteville shale of Arkansas and the Woodford shales of Oklahoma, are far smaller than the Appalachian Basin

---

* The industry uses a 6:1 ratio (or 5.8:1 or similar ratios) to convert gas measures into barrels of oil equivalent. One thousand cubic feet of gas (at a specified temperature and pressure) has the same energy value as a sixth of a barrel of oil. It is at best roughly right, since various oils and gasses, depending on their sources and processing, have quite different energy values.

shales, with a combined area of about 13,000 square miles. The depths are quite variable, ranging from 1,000 feet all the way down to 14,500 feet, or nearly three miles, in a branch of the Woodford. A 2010 survey of most of the area by the USGS concluded that there are total recoverable reserves of thirty-six Tcf of gas, or about the same as the Utica.[9]

The Gulf Coast region has three major formations, with an area of about 14,000 square miles between them. The Haynesville shale in east Texas and Louisiana, according to industry sources, has very large gas resources, while the Eagle Ford in east Texas has been producing large quantities of shale gas and "tight" oil, or oil retrieved from oil-bearing shale and sandstone. Industry sources believe that its oil production capacity is of the same scale as in the Bakken formations (see below). The Floyd-Neal shales in Alabama and Mississippi are also believed to have substantial reserves of gas.

The producing areas of the Southwest region include the Barnett shale around Dallas-Fort Worth, the location first exploited by George Mitchell; the Permian Basin region in far-west Texas, once the world's most productive oil region; and further west into the Avalon-Bone Springs region of west Texas and New Mexico. The Barnett is the most intensely developed shale gas and liquids site in the world, with rigs that once dotted the Dallas-Fort Worth airport. Production began with natural gas, but over the past year has swung sharply toward liquids, since the heavier fractions like ethane, which competes with petroleum-sourced naphtha as a chemical feedstock, command higher prices than methane. As this book goes to press, Texas producers are enthusing over what may be a massive find in the state's Permian Basin, which is plausible, given its once-great oil endowment.

The Rocky Mountain region has two distinct producing areas—the Green River and Piceance Basins of Wyoming and Colorado; and the Bakken oil formations of North Dakota and Montana, and well into Canada, which may be the single largest oil producing formation in US history.

The Bakken formation mostly consists of oil-bearing sandstone squeezed between two shale formations. Fracturing and well technologies are basically the same as for natural gas, adjusted for the different weight and viscosity of the material. Production has jumped from 10,000 to 15,000 barrels a day in 2008 to a half million a day in 2012; ExxonMobil expects the Bakken to reach a steady state of a million barrels a day within the next several years.[10] North Dakota is now the second largest oil-producing state after Alaska.

Wyoming and Colorado have rich natural gas and NGL shales. A 2002 USGS assessment of the Green River Basin in Wyoming concluded that there was more natural gas than even in the Marcellus, and it is found in relatively porous sandy formations that are accessible by vertical wells. But the area is still better known for its spectacular reserves of oil shale, which may be as much as 60 percent of the world total, although it will be a long time, if ever, before they make a meaningful contribution to the usable hydrocarbon supply.

The final region, the Far West, so far consists of a single oil-bearing area—the Monterey/Santos shale formations in southern California. California was once one of the largest of America's oil-producing regions, and preliminary assessments are that the oil-bearing shale formations may hold far more oil than even the Bakken.

The Players

The table on page 35 offers a snapshot of companies that are competing in the shale oil and gas production process.* I have left out the oil majors—ExxonMobil, Shell, Occidental, Hess—who are rapidly building their footprint in the business. Although ExxonMobil and Shell are among the largest players in the shale and related industries, their other operations are so big that their their financial reports necessarily provide less detail than those of more mono-focused companies do. The companies in the list have varied histories. Several entered from backgrounds in technically challenging offshore drilling, others previously had pipeline, oil field servicing, or gas processing businesses, and still others were pure startups by experienced people. All of them are substantial firms; only one on the list had annual sales of less than $1 billion.

Over the past two years, the dominant factor in the American industry has been overproduction—not so much with respect to demand but with respect to transport capacity from the production areas. Production of natural gas has run well ahead of "out-take" capacity, so prices of natural gas have fallen very sharply—from as much as $8 per thousand cubic feet (Mcf) to, at one point, slightly under $2/Mcf last year. As this is written, prices are hovering in the $3.50/Mcf range, close to the $4–6/Mcf the industry needs to be profitable. A

---

*The ten listed are drawn from throughout the top twenty names in a shale drilling exchange-traded fund (ETF). Information is from company reports; financial information is from the most recently filed 10-Ks.

COMPANY	SALES $MMS	NET INC. $MM	EMPLOYEES	COMMENT
Chesapeake Energy	11,635	1,570	12,600	Entered industry in 1989, concentrating on shale gas, NGLs, and oil. Positions in all major shale formations.
Devon Energy Co	11,454	4,704	5,200	Strong positions in Texas & Oklahoma—Barnett, Permian, Cana-Woodford, Wyoming & Colorado, Ohio Utica.
Eog Resources	10,126	1,091	2,550	Focus on crude production: Eagle Ford, Barnett, Louisiana, Bakken.
Encana Corp	8,467	128	5,000	NG producer, shifting rapidly to liquids. Based in Alberta. In US, is active through the Rockies to Texas and Louisiana.
Noble Energy Inc	3,763	453	1,876	Deepwater in Gulf; offshore in West Africa. In US, positions in wet areas of Marcellus, and in tight oil in Colorado.
Crescent Point Energy Corp	1,745	201	NA	Canadian company gas, NGLs, and oil in Calgary and Alberta and Canadian and US Bakken. Expanding into Colorado.
Eqt Corp	1,640	480	1,835	Concentrated on Marcellus, with additional wells in the Devonian formations in the Appalachian Basin.
Range Resources	1,173	58	756	First company in the Marcellus shale. Strong Appalachian positions. Also in southwest in Permian and Anadarko Basin.
Cabot Oil & Gas	848	122	529	Concentrated on Marcellus for gas; and Texas-Oklahoma for liquids, primarily in Eagle Ford.

similar, but less severe, situation exists with shale oil production. Under normal circumstances, there is a tight relationship
between American prices for crude based on a "sweet" non-
sulphurous crude pricing hub—the West Texas Intermediate
(WTI) and the "Brent" price for the more sour oil from the
Arabian Peninsula. Because of a glut of American crude that
has overwhelmed shipping capacity, the WTI has been consistently 20 percent or more lower than the Brent.

The industry continues to invest aggressively. Capital
spending in 2013 is projected to be $348 billion, with about
two-thirds of it in the upstream—land acquisition, exploration, and production. The investment announcements
highlight the growing dominance of the industry majors;
ExxonMobil, Shell, Hess, Chevron, and Conoco Phillips are
all up at the top of the list. The biggest jump in spending is
the near-quintupling of investment in pipelines. Gas prices
will not rise to profitable levels until the industry can reach
more of its natural customers; ratcheting up pipeline investment is the only way to make that happen.

So How Much Do We Really Have?

In his January 2012 State of the Union address, President
Obama said that "We have a supply of natural gas that can
last America nearly one hundred years." That sounds like a
definitive report of a measured quantity of gas. In fact, it is
highly speculative. Only a tiny fraction of the known natural gas shale formations have had wells sunk in them, so
there are few data points on recoveries, gas quality, flow rates,
and other parameters. All current estimates are drawn from

mathematical models that at their best can make only gross approximations. The USGS uses a probabilistic model that gathers the standard modeling inputs, the experience in similar geologic formations, the percentage of the target area that has *not* been tested, as well as the constraints of current technology. The rapid progress of "current technology" by itself has generated steady increases in the estimates of recoverables.

Until last year, that is. At the time of the president's speech, the US Energy Information Administration's (EIA) best estimate of technically recoverable resources* (TRR) of natural gas was 750 Tcf, the rough energy equivalent of 130 billion barrels of oil, which is about thirteen years' worth of current production by Saudi Arabia—so, yes, that is a lot of energy. But in the summer of that same year, the EIA reduced its gas recovery estimate to just 482 Tcf, due entirely to a two-thirds reduction in the estimate of reserves in the Marcellus shale.

The backstory is that in 2002, the USGS surveyed the Marcellus and came up with a mean probability reserve of only 1.9 Tcf of gas, based on the performance of older vertical wells. When the advent of horizontal drilling brought a rush on gas extraction by entrepreneurs into the Marcellus, production quickly got on a track to exceed the USGS's estimate within just a couple of years. The EIA asked for another survey, which was not completed and published until

---

* The EIA uses two definitions of inground supplies of energy—proved reserves and technically recoverable resources (TRRs). Proved reserves are those that the data suggest "with reasonable certainty" can be recovered from known reservoirs "under existing economic and operating conditions." TRRs add resources that can be recovered with current technology *without* regard to economic or operating conditions.

mid-2011. By then the EIA, based on its own work and that
of consultants, had published a TRR estimate of 410 Tcf. It
was an embarrassment therefore when the USGS, who are the
science guys, came in with a mean estimate of 84 Tcf TRR, a
big number, but not even a fourth of what the EIA had been
carrying. The EIA massaged the USGS figures a bit—USGS
had still not assumed the availability of horizontal drilling
in some areas—but followed their primary reductions, which
were based in part on lower quality assessments of large areas
of the Marcellus, producing a final number of 141 Tcf.[11]

A second disquieting sign is that early wells appear to have
a more rapid depletion rate than industry and EIA projections
assume. The unconventional hydrocarbon industry is still so
new that only a relative handful of wells have as yet produced
much in the way of useful data. The Texas Barnett shale has
the longest track record. An *Oil & Gas Journal* analysis published in late 2012 found only two shale gas plays—the Barnett and the Fayetteville—with meaningful data, along with
one tight oil field—the Elm Coulee in the Bakken play.[12] All
three had started with high production rates but experienced
rapid declines. Companies love the high initial production
since they recover costs faster, but long-term recovery estimates and capitalization values assume production schedules
of twenty-five to thirty years. If the few early decline rates
turn out to be representative, total recoveries would be only
about half of the current estimates. Combining that with the
recent TRR reductions by the EIA suggests that the energy
boom would be of far shorter duration than currently imag-
ined—twenty or thirty years perhaps, or even less.

While disappointments are not impossible, those data are
only straws in a sandstorm, and it's far too soon to attach

much significance to them. Unconventional gas and oil exploitation is a brand new industry that has already run far ahead of the expectations of even a decade ago, and only a tiny percentage of potential drilling opportunities have been actively explored. Witness the several new announcements over the past year. Drilling technology is still evolving rapidly, and there is no warrant for assuming that early recovery efficiencies won't be improved upon, or that the engineers won't learn to exploit areas that appear relatively unattractive today. The EIA and other forecasters freely concede that their forecasts are the grossest of approximations that will necessarily require frequent large revisions. Perhaps the best indication of the reality of the opportunity is that the oil majors—ExxonMobil, Shell, Chevron, Marathon, and others—are jumping in with both feet, not only starting their own production sites but buying up ones already developed by smaller companies. They are also champing at the bit for permits to spend up to $100 billion or even more to build the massive projects required to support the export of natural gas (see Chapter 2). The oil majors are among the most experienced energy companies in the world, and they have clearly sized up the opportunity and made their own very positive assessments. While the growing presence of the majors may raise concerns about future cartelization of the industry, it eases worries that it is merely a swamp-gas will-o'-the-wisp.

The most recent research, published just as this book was going to press, is an impressive new study from the Bureau of Economic Geology at the University of Texas at Austin. An assessment team collected complete production and geologic records from all 15,000+ wells that had been sunk on the Barnett shale by the end of 2010. Combining the actual production records with a new geology-based production model to project the output of new Barnett wells, the team concluded that actual production would be in a mid-range between the most optimistic and most pessimistic assessments—in other words, much like current EIA forecasts. Further, they scotch fears expressed by some that all the "sweet spots," have already been tapped. At least in the Barnett, it looks like there is a lot of good rock left. All of it is bracing news for the industry. By end of 2013, the team will extend the study to several other shales, including the critically important Marcellus.[13]

But what if the worst case scenario comes true, and the shale boom lasts only twenty or thirty years? The golden age of American manufacturing is often placed in the very prosperous postwar period of about 1948–1968—only twenty years. We shouldn't turn up our noses at another twenty-year run like that.

THE "OTHER" UNCONVENTIONAL HYDROCARBONS

Coalbed Methane

Large amounts of methane are trapped in coalbeds. Coal is porous and adsorbs (attracts) methane. Since concentrated methane is explosive, miners long ago developed methods for draining the methane from mines, often by drilling vent holes. The large-scale capture and reuse of the methane is a more recent development, at least outside of Germany, which began capturing coalbed methane some forty years ago. Australia, China, and the United States are now all extracting coalbed methane in commercial quantities. Unlike shale gas, coalbed methane contains few liquids, so its production has shrunk with the current low prices for dry gas. There are multiple techniques for extracting methane from coalbeds, some of which involve fracturing and lateral well drilling, much as in shale gas operations, although the majority use other methods, which may involve vacuum pumps and boreholes, and occasionally even blasting. Unlike shale plays, extracting methane from coal mines usually *produces* water. Water builds up in coal seams, and must be cleared to release the methane. Water recovered from coalbeds is typically highly toxic and requires careful handling. Coalbed methane production, at about 2 Tcf annually, accounts for about 10 percent of US dry gas production. The EIA forecasts that its production will fall slowly through 2035.[14]

Oil Sands

Oil sands are sand formations infused with bitumen (oil tar) a heavy, extremely viscous hydrocarbon that does not flow at normal temperatures. Oil sands are widely distributed and represent a potentially vast store of energy. Currently, the major development is concentrated in the Alberta province in western Canada. Shallow oil sand deposits are removed by strip mining techniques (using some of the largest power shovels in the world) and treated in separation centers using water and caustic chemicals to separate the bitumen from sand and other waste. Deeper deposits are treated in situ, by pumping superheated steam and chemicals into the sands, melting the bitumen, so it can flow out from the sands. Bitumen-derived oil is heavy and sour, but with additional processing, it can be converted into the light, sweet, oil preferred for transportation and other big-market applications. But

extraction and processing, which often involves mixing the processed bitumen with conventional petroleum products, is expensive and energy-intensive, while the affinity of bitumen for heavy
metals and other toxic substances imposes complex and expensive treatment issues. Alberta produces about 1.5 million barrels of
high-quality oil a day. Production levels may not be sustainable in
the face of a glut of Canadian oil, due to a backup in transportation
facilities. The controversial Keystone XL pipeline, held up by the
Obama administration, is key to ending the glut, by moving the oil
to Gulf Coast export facilities. The cost of oil sands production may
rise to unacceptable levels as the industry moves beyond the most
readily accessible production fields.[15]

Oil Shale

Oil shale is a sedimentary rock in which rock-like particles of kerogen—the source organic hydrocarbon—is bound tightly within
quartz, limestone, or other rock forms. (The industry uses the term
"tight oil" to distinguish *oil-bearing* shale or sandstone from "oil
shale.") Conceivably, it is the most abundant, if one of the least accessible, forms of hydrocarbon. Rough estimates are that there are
some three trillion oil-equivalent barrels of oil shale in the world
(some estimates are much higher); two-thirds are in the United
States, primarily in the Green River and Piceance Basin formations
in the Wyoming and Colorado Rockies, with a substantial portion
of them on federal lands. The United States authorized the com-
mencement of several research, development, and demonstration projects during the George W. Bush administration, which
have since been revised under the Obama administration. Two
awards have been made on 160-acre demonstration tracts, one to
ExxonMobil and one to National Soda Inc., a Colorado firm. Each

of them will use in situ methods, which are viewed as more en-vironmentally sound than mining the rocks for reprocessing in surface plants. The two are using different processes, but each involves heating the subsurface target rock until it reaches a liq-uid state and then separating the kerogen with techniques that are analogous to, but more complex than, typical unconventional technologies. Production from oil shale is either not included at all or assumed to produce little usable product in the major energy near-term forecasting programs.[16]

CHAPTER 2

The Return of the Whales

N ucor, America's biggest steel company by market value, has transformed the steel industry by focusing on "mini-mill" steel production, using scrap steel as its raw material. Mini-mills once actually were small steel mills designed for economic production of customized orders, often using more advanced technology than the industry giants. Nucor seized on the mini-mill model of flexible, low-cost, high-quality steelmaking and drove it to automobile-production scale. Last year, its US production was nearly thirty million tons from twenty-three mills, and it is the dominant American structural steel provider, with profit ratios double those of the big traditional companies. It is also one of the very few large employers that made no layoffs during the financial crash, and it still operated profitably throughout the downcycle.[1]

Its continued growth, however, has been threatened by rising prices of quality scrap steel as more and more companies adopted Nucor's technology, and so it set a company goal of finding seven million tons of high-quality, affordable, scrap substitutes. Nucor could buy and convert pig iron to steel, but was uneasy with the volatility of merchant pig prices. Building blast furnaces to smelt its own iron from ore was another alternative, but to be cost-effective, blast furnaces must be very large and operate more or less continuously, while Nucor's business strategy is all about fast adaptation and flexible production schedules. Blast furnaces also match up poorly with Nucor's image as one of the world's "greenest" steel companies. Another solution was "direct reduced iron" (DRI) a method of melting iron ore and separating out the slag in an electric furnace similar to the ones used to melt scrap. But melting ore requires a lot of energy, and since the preferred DRI processes are gas-fueled, high American gas prices made the idea impractical. Until now.

In late 2012, Nucor announced a long-term investment and supply deal with Encana, a Canadian shale player that Nucor credits with attitudes similar to its own toward safety, employees, and the environment. Nucor committed to spend $3.6 billion to capitalize Encana's expansion of its natural gas properties in return for a 50 percent working interest in the resulting wells. Normally, a 50 percent working interest arrangement confers the right to take half the profits from the wells, but in this instance Nucor will take half the gas at a cost-based price. The volumes contemplated should give Nucor sufficient gas to fuel two large DRI plants in Louisiana—the first of which will open this year—plus all its steel operations. Nucor is a global company with operations around the world, but has so far maintained almost all of its manufacturing in the United States, and the DRI strategy will be a big factor in allowing them to stay that course.

The benefit for Encana is at least as great. Encana's initial development was tilted emphatically toward shale gas, both in Canada and the United States. But the temporary glut in American natural gas that drove prices to unheard-of lows (in 2012 down to less than $2 per thousand cubic feet [Mcf], which hardly meets the cost of production; on a barrel of oil equivalent basis, it is close to $12 crude) required that the company's oil and gas assets be written down to reflect the fall in their value. Gas-heavy companies have taken big hits to their asset positions. Those are non-cash charges, but banks don't lend open-handedly to companies with declining net worth, especially if the companies are aiming to expand gas production. Like other exploration and production (E&P) companies, Encana has been switching its production to liquids, but its holdings are not as liquid-rich as those of some others. The Nucor deal has given it the wherewithal to maintain an active drilling program in the Colorado's promising Piceance Basin.

The deal may well be a prototype for other companies. Fears of yo-yoing gas prices have a real basis, and gas investors have been battered over the past couple of years, although as this is written in early 2013, prices have recovered to about $3.50, closer to the $4–$6 range that most observers feel is a healthy median. The very low recent prices should be a temporary phenomenon, caused by the mismatch between the gas production boom and the limited transport capacity out of the shale regions. Many producers have shut down their gas wells—they are fairly easy to restart—and are supplying their customers from stored gas. As the gas glut is worked down and the multiple new pipeline projects come on line, prices should stabilize. But it may be a while before investors can be enticed to take a new plunge into gas plays.

The Nucor story is just one example of the spinoff effects of abundant, cheaper energy. A number of major industries—aluminum, chemicals, iron and steel, paper—have energy costs that are significant percentages of their sales. In aluminum and chemicals, energy consumes 5–6 percent of total revenues. US chemicals production is about $1.5 trillion, so a wholesale shift to much less expensive natural gas would produce tens of billions of dollars of profits.

The impact of cheaper energy on American manufacturing will be all the more powerful since it comes on top of a broader momentum to bring much of our offshored production back home for other reasons—"reshoring" is a voguish term at the business schools. Peoples throughout south and east Asia are widely achieving middle class lifestyles, and they want better pay and benefits, shorter workweeks, and some of the other perquisites that their Western peers enjoy. The "low-cost" label customarily applied to places like China and India is out-of-date.

At the same time, American production costs have mostly been falling, especially after the ruthless cost-cutting that companies employed after the 2008 crash. America may never produce hard goods as cheaply as China, but the cost convergence spotlights the embedded inefficiencies of offshoring. Inventory controls are compromised when goods spend weeks on ships. Distant contracted production necessarily increases the challenge of quality control—it's harder to make rapid design changes or to react to customer complaints. A big American energy cost advantage can be the last piece of the equation that tilts the decision toward reshoring.

The New Energy Economy

To begin with, energy by itself generates lots of jobs without taking account of the productivity jolt it gives to other industries. IHS is a global consulting firm with a major practice in energy. (Daniel Yergin, one of the world's leading academic authorities on energy is a vice chairman of the company.) In the fall of 2012, the company published an analysis of the economics of the unconventional oil and gas sectors using models developed by IHS Global Insight, their country intelligence and economic forecasting business.[2]

Their estimates of the employment impact are in Table 2-1. Those numbers are built from three estimates: first, the direct employment in the unconventional energy industry—geologists, rig and equipment operators, mechanics, laborers, and the like; second, the employment in their supply chain—a typical well requires up to one hundred tons of high-quality steel pipe, fleets of trucks and trailers, a small hangar full of earthmoving, drilling, and other equipment, some very high-end seismic and imaging gear, as well as chemicals, special sands and other supplies; and finally the "induced employment" generated by the new spending by both the direct and supply chain workers. The numbers in Table 2-1 are premised on the assumption of a steady expansion of unconventional production, and a rising pace of opening new wells—this in

TABLE 2-1: Unconventional Oil & Gas Employment

	2012	2015	2020	2035
Total	1,749,000	2,511,000	2,985,000	3,499,000
Direct	360,000	506,000	600,000	724,000
Supply-chain	538,000	770,000	916,000	1,074,000
Induced	850,000	1,234,000	1,469,000	1,700,000

an industry that, for all practical purposes, didn't exist a decade or so ago.

Note that supply-chain employment growth is greater than the growth in direct employment. One of the bullish features of the industry is that for the moment, it is a US specialty, and the technology and supplier know-how has a pronounced made-in-America stamp. China has very large shale deposits that may be more technically challenging than the primary American sites, and while its government is anxious to jump-start a home shale industry, they freely concede that they will need Western help to do so. In 2009, the United States and China launched the *US–China Shale Gas Resource Initiative* to ensure close contact as the industries evolve. Chinese companies have been permitted by both governments to take minority positions in American companies, and a few Western companies, like Schlumberger, the giant oil field servicer, have been acquiring stakes in Chinese firms. Schlumberger, technically a Curaçao firm, is a truly global company with principal offices in Paris and The Hague, but with its primary headquarters in Houston. It is a major operator of US unconventional drilling and production sites. Necessarily, the suppliers for that business are primarily American companies. China, as always, will be anxious to insource its supply

TABLE 2-2: Annual Upstream Capital Expenditure

CURRENT $Bs	2012	2015	2020	2035	2012–2035*
Drilling	28.0	41.5	57.7	122.4	1,761.0
Completion	46.9	67.2	92.3	188.3	2,737.4
Facilities	6.7	9.6	12.6	24.5	370.7
Gathering System	5.7	8.0	9.9	17.9	279.3
Total	87.3	126.3	172.5	353.1	5,148.4

* Cumulative spending over the entire period.

requirements as soon as possible, and will doubtless quickly supply its own pipes and excavators, but insourcing the most sophisticated equipment and the skills to use it is likely to take a decade or more.[3]

Also note that IHS analysts calculated just the employment from *upstream* components of the unconventional energy rollout over the next couple of decades. Upstream processes include drilling, completion (fracturing and starting the upflow of oil or gas), facilities construction, and gathering (sending the product via pipeline to a local hub). Table 2-2 estimates upstream capital spending.

Midstream processes include pipeline and other transportation from local hubs and usually further processing, while downstream processes cover the complex refinery processes, and marketing and transportation to end customers.

All oil products, and most NGLs (natural gas liquids) and other condensates require refining, usually by catalytic cracking. The United States had substantial excess refining capacity when the unconventional boom got underway, left over from the days when it was one the world's largest conventional oil producers. That is now being used up and requires substantial modernization. Very pure dry gas, of the kind normally found in the Marcellus shale, often requires only a washing and filtering before it is "pipeline ready," which is usually handled in the upstream phase. Exporting gas, however, will require multi-billion-dollar liquefaction facilities.

Transportation is typically by pipeline. The United States already has an extensive pipeline network, but industry experts suggest that for the foreseeable future it must add at least 2,000 miles a year both to accommodate product volume, and to make connections with new producing areas. The ongoing imbalance between production and transportation is behind a small economic tragedy unfolding at the Bakken shale in North Dakota and Montana. The Bakken is primarily an oil play, and its gathering and transportation systems are oriented to petroleum refineries. But its oil deposits typically come with large quantities of "associated" natural gas. Since the Bakken lacks gas transport and processing capacity, it is flared.*

Wages in the industry are quite good by today's standards. First-line supervisors and skilled and semiskilled workers at the rig sites are typically paid in the $20–$33 per hour range. Overtime that is built into the standard work schedules has resulted in very young workers taking home gross pay in the $100,000 a year range, with better than average benefits.

---

* Flaring is more environmentally friendly than simply dumping the gas into the atmosphere, since it converts methane into CO_2, which has less of a greenhouse effect than methane, at least in the short term. Concentrated quantities of methane are also explosive.

There is a downside to the good pay, since the drilling sites are often in quite remote parts of the country. The boom in thinly populated North Dakota has resulted in large temporary settlements of single men with associated problems of alcohol, violence, and sexual assaults. Companies appear to be addressing such issues haphazardly at best. By no means are all wells in remote areas, however. Some of the most productive wells in the country are spotted throughout the Dallas-Fort Worth airport, and virtually surround Fort Worth to the west. The lateral from one well is said to terminate under the Texas Christian University football stadium. The workers at the Devon sites, I was told, nearly all lived within reasonable driving distance of their work location.

The View from the Chemical Industry

Organic chemistry deals with carbon-based compounds, usually with carbon-hydrogen bonds—the same hydrocarbons that we use for fuel. The industry's base products include plastics, vinyl chloride, polyethylene, and poly vinyl chloride (PVC), among others; its end-use products include Styrofoam, tires, sealants, adhesives, films, liquid crystal screens, and a vast range of fibers like nylon and polyesters—in short, a huge portion of the things that we depend on for daily living. The organic chemical industry is therefore a massive user of hydrocarbons, first as the raw material, or feedstock, for its plants, and second to generate the heat and pressures for the giant "crackers" that break the hydrocarbon molecules into smaller chains for reassembly. It consumes hydrocarbons for feedstock and fuel in roughly equal volumes.[4]

Natural gas was abundant in America even after American oil production began its swoon in the 1970s, and the feedstock favored by American chemical plants was ethane, the lightest of the natural gas liquids (it's usually a gas in the atmosphere). The European industry favored naphtha, derived from petroleum. It is a more complex compound, but can be cracked and reassembled into the same products that Americans constructed from ethane. As a general rule, the American industry was very competitive so long as the price of a barrel of oil was about seven times the price of natural gas at the Henry Hub, a gas distribution focal point in Louisiana that is the pricing point for natural gas futures contracts.*

The American industry did very well in the first part of the 1970s. The country still had plenty of gas, and oil prices were soaring—to more than twenty-two times the price of gas. But as gas supplies dwindled, the cost advantage steadily eroded. American plants made a brief recovery in the 1990s, due to the oil price disruptions of the Gulf War, but when the government encouraged electrical utilities to shift to gas through the later 1990s, gas prices jumped with devastating effect on the chemical industry—until shale gas burst on the scene in volume around 2002. Over the last several years, the industry has been recovering strongly. By 2010, the United States was one of the most competitive countries in the world for organic chemicals, with plants running full, and exports up and rising.

---

* A barrel of oil, remember, has about six times the energy content of the thousand cubic feet of gas which is the pricing unit at the Henry Hub. The 7:1 rule of thumb takes into account that ethane is somewhat harder to transport and to convert into ethylene than naphtha.

Economists at the American Chemistry Council (ACC) have generalized from their industry's experience to assess the effect of the shale gas bounty on eight energy-intensive industries.[5] Ranked in order of total energy usage, they are chemicals (excluding pharmaceuticals), plastic and rubber products, fabricated metal products, iron and steel, paper, aluminum, and foundries.

Using econometric models similar to those used by the IHS analysts, the ACC's analysts projected the additional annual employment and GDP generated by lower fuel prices by 2017:

TABLE 2-3: Economic Impact in Eight Industries

	EMPLOYMENT	PAYROLL $Bs	OUTPUT $Bs
Direct	200,000	14.6	121.0
Supply-chain	462,000	31.7	143.8
Induced	517,000	24.2	76.8
Total	1,179,000	70.5	341.6

Those are serious numbers. The $342 billion total output effect is about 2 percent of GDP for the United States. And note that these data comprise only these eight industries, so they omit the gains in the energy industry itself, which was the focus of the IHS study. In normal times, rapid growth in one sector would be managed in substantial part by shifting workers and resources from other, less favored sectors. But the Great Recession has left so much slack in the economy, most of this should be additive.

There are a number of other analyses that coalesce around similar numbers. A detailed 2012 Citigroup study suggested that the cumulative effect of the energy revolution would be 2.2–3.6 million net new jobs by 2020.[6] That is quite close to the combined estimates of the IHS and ACC studies, which between them covered the economic impact from the growth in both the energy and biggest energy-using industries. (At the high end, that would increase total employment by about 2.5 percent.) Dow Chemical has recently listed 108 major industrial projects planned, announced, or already underway by energy-intensive manufacturers, thirty-three of them with 2012 or 2013 start dates and sixty-two with start dates by 2015. The total planned investment, just in these projects, is estimated at $95 billion.[7]

Activity on the ground reported by the consultancy PWC supports the bullishness of the forecasts. Dow itself restarted a long-mothballed ethylene plant unit in Louisiana in January of 2013; the company is also planning to open a new ethylene unit on the Gulf Coast in 2017, to build out more capacity in existing facilities in Louisiana and Texas, and to build a new Texas propylene plant, all in the near term. Formosa Plastics is planning a $1.5 billion expansion program in Texas, and West-lake Chemical is expanding its ethylene capacity in Louisiana and considering another unit in Kentucky. Chevron Phillips and Bayer are scouting sites for new ethylene capacity both on the Gulf Coast and in the Marcellus. Shell has settled on the site for a major new ethylene plant near Pittsburgh. US Steel, the pipemaker TMK IPSCO, and the French steel company Vallourec, are adding capacity and building new plants in the shale regions to supply the pipe and other equipment needed by the producers, as is Voestalpine, a major Austrian steel-maker. US Steel is also exploring substituting natural gas for coke in some of its processes, which it estimates could save it about 1 percent of its production costs—no small amount in the tightly competitive global steel game. Perhaps the biggest commitment of all is the announcement that SASol, a South African specialist in gas-to-liquids (GTL) conversion, a tricky and controversial method of converting lighter gases into higher value products like diesel and gasoline, is planning a $14 billion GTL facility in Louisiana. GTL is an expensive process, and SASol will require a long period of low gas prices to make the investment pay off, but the company seems confident it can pull it off.[8]

The Rest of Manufacturing

Over the past two years, the Boston Consulting Group (BCG) has published a series of studies on the emerging recovery of American manufacturing. Although they count falling US energy prices as a positive factor, they are not the linchpin of the thesis. BCG's argument, rather, is that productivity-adjusted Chinese and American labor costs are rapidly converging, even as the US matches up well against China in the costs of many other factors of production, like land, transportation, and the cost of equipment. By 2015 or so, BCG estimates, an American company will save only about 10–15 percent of costs by manufacturing a kitchen appliance in China, which is too little to justify the long delivery lead times and other aggravations that come with offshoring.[9]

China's rise as a global manufacturing power has been extraordinary. In just the nine years from 2000, China's exports leaped fivefold, and its share of global exports increased by two and a half times. And it happened nearly across the board, in a wide range of industries. Its share of global exports in textiles, furniture, ships, telecom equipment, office machines, and computers all rose into the 20–33 percent range. The United States clearly bore the brunt of that onslaught. In the first decade of the 2000s, America lost 6 million manufacturing jobs—in the mild recession year of 2001, the country lost 1.5 million manufacturing jobs, far more than even the 900,000 jobs lost in 2009, the worst year of the Great Recession. The reasons the initial impact was so severe include, among others, America's commitment to a normal trade relation with China,[10] the convenient sea-lanes and the well-developed West coast container ports, and the binge of consumer credit creation by the Greenspan Federal Reserve. By the midpoint of the decade, however, it was clear that Chinese exporting prowess was being felt everywhere—in western Europe, in the rest of Asia, in Latin America, and in Africa.

But as a wise man once said, unsustainable trends tend not to be sustained. China's great success and the extraordinary achievements of its workers in moving up the technology curve are turning it into a middle-class nation. Chinese worker productivity is still growing at about 8 percent a year, an extraordinary rate, but worker compensation is growing more than twice as fast. In 2000, average worker compensation in the Yangtze delta, one of China's most dynamic manufacturing areas, was about $0.72 an hour. By 2010, it had risen to $8.62, adjusted for productivity. That's still a lot less than the $21.25 averaged by manufacturing workers in the American South in 2010. But labor is only a part, often a small part, of the total cost of product, and when you consider

America's competitiveness in so many other factors of production, a modest wage differential is not enough to drive business offshore. The BCG analysts project that, by 2015, total wages in the Yangtze delta will be up to $15.03 an hour compared to $24.81 in the American South—which is parity, for all practical purposes. Wages in India are on a similar course. Other Asian countries further down the development curve, like Vietnam and Bangladesh, have very low wages, but are not nearly as sophisticated or as productive as China, and are unlikely to repeat the Chinese experience, at least not in so compressed a time.

The loss of American manufacturing has never been as catastrophic as it sometimes feels. At the end of World War II, the United States accounted for some 40 percent of global manufacturing production. From such lofty heights, there was nowhere to go but down. It took a generation for Europe and Asia to recover from the war's devastation, while the United States grew fat and lazy as the world's richest-ever hegemon. That era came to an end when the Germans, Japanese, and Koreans launched an all-out assault on American automobile, consumer electronics, and computer and chip markets. After a violent restructuring in the 1980s, the United States emerged as the market leader in most high-value-added industries, like microchips, aerospace, and pharmaceuticals. By the close of the 1990s, it was once more the global "hyperpower," before it was kneecapped by inconclusive foreign wars, the Chinese trade onslaught, and the housing bubble and related misdeeds of its financial sector. Despite all that, inflation-adjusted American value-add in manufacturing actually increased by a third to $1.65 trillion between 1997 and 2008, just before the recession. The collapse in US manufacturing was in *employ-ment*, not in output, but falling employment is always the flip side of a highly productive industry.

Among advanced countries, the United States was clearly an outlier in the degree that its workers bore the brunt of the financial crash. One of the anomalies of the recession was how corporate profits held up even as sales plummeted. Lavish federal bailouts made 2009 one of Wall Street's most profitable years, while the ordinary people who lost their jobs and homes by the millions could fall back only on the least generous social safety net among the major industrial countries. Companies laid off workers well ahead of sales downturns, and single-mindedly focused expenditures on labor-saving equipment. Unattractive and unfair as the process was, the restructuring has made the United States one of the most desirable manufacturing sites in the world, especially in states like Virginia, Tennessee, Georgia, the Carolinas, and Alabama, where land, labor, and energy is cheap and factory productivity is very high. Just within the past year, Japanese and Korean automobile companies have been adding to their American plants, or building new ones, in order to service the European market; Siemens has begun producing gas turbines in Virginia for export to Saudi Arabia; and Rolls-Royce has opened an airplane parts plant to serve its global customers.

The BCG analysts estimate that the United States' total cost advantage over Japan and Europe is in the 25–45 percent range, even as it offers one of the world's best trade logistics capabilities.* They also identify seven American industries that

---

*There are worries about the availability of qualified American workers, but they may be overdrawn. The energy industry has no obvious recruitment issues. If you pay them, it seems, they will come.

are on the economic tipping point of a substantial reshoring. They are: computers and electronics; appliances and electrical equipment; machinery; furniture; fabricated metals; plastics and rubber; and transportation goods (meaning primarily the components industry). Collectively, they account for well over 10 percent of American consumption. The reshoring opportunity, BCG estimates, coupled with the parallel gains in exports, should shave $100 billion off the trade deficit by 2015—in addition to the shrinking deficit in energy trade. The "global imbalances" so deplored by global financial authorities over the past decade or so may shrink away far faster than anyone expected.

The case for an American manufacturing renaissance, therefore, is not dependent on the new availability of inexpensive fuel and feedstock. A major revival is quite likely in the works anyway, but the energy bonus will help enormously in getting it off the ground.

The Siren Call of Exports

How much we actually realize those benefits, however, may turn, at least in part, on a question that has shadowed all of the economic projections: Assuming the energy boom is real, can America maintain its energy price advantage? The industry likes to think of the world's energy supply as a bathtub—you take something out or put something in, and the level in the whole tub rises or falls. But at the moment at least, natural gas clearly doesn't work that way. North America currently has a gas glut and rock-bottom prices. Europe gets most of its gas from Russia, which prices it by reference to the cost of oil, or about triple what Americans pay for gas. East Asia, especially

Japan and Korea, pony up four or five times American prices; they are energy-poor—the Fukushima disaster has left Japan in desperate straits—and they must be supplied by tanker, which is expensive.

The US gas industry has mounted an all-court press to convince the government to allow the export of natural gas. Economic theory is on their side. Prices set by willing sellers and willing buyers in open markets are always the fairest and the most market-efficient. So if energy product could move seamlessly to the best market, quantities of gas and of petro-leum with the same energy values would be priced approxi-mately the same—the price of gas and the price of oil would become linked. The increase in efficiency would allow all the world's consumers to pay slightly less for energy. American gas companies, and especially the largest players like Exxon-Mobil and Shell that are best positioned to trade globally, would get a nice profit jump. But achieving a higher state of *global* efficiency would not do much for the American eco-nomic revival that the oil and gas industry has been trumpet-ing as their gift to the nation.

It's a complicated issue, and I will try to unpack the main questions here. By law, companies must secure permits from the Department of Energy (DOE) before exporting natu-ral gas. Export permits may not be denied for sales to coun-tries with a free trade agreement (FTA), of which there are only three with a major interest in importing gas. Canada and Mexico, which are covered by President Clinton's 1994 NAFTA treaty, have long had gas pipeline connections with the United States, with flows in both directions. The newest free trade partner, South Korea, is eager to tap into cheap

American gas. As of this writing, twenty-three companies have filed gas export permit requests, only one of which, from Cheniere Energy, has been approved for general exporting from a facility that will go into operation in 2015 on the Gulf Coast of Louisiana.[11]

It is not easy to export natural gas. It must first be highly purified and then liquefied (LNG for liquefied natural gas). At negative 162°C, natural gas becomes a liquid with about 1/600 the volume of normal gas, and an energy density close enough to conventional oil products that it is cost-effective to ship. The importing country requires a re-gasification plant of similar complexity. LNG plants and re-gasification plants, both cost billions and take years to build. In the late 1990s, as American sources of natural gas dried up, prices spiked above $60/Mcf, and American companies rushed to build re-gasification plants to *import* LNG. Forty-seven plants were approved and a number were built at a total cost estimated in the $100 billion range. All of them were mothballed by the shale gas boom. The Cheniere LNG plant will be a converted re-gasification facility; much of the infrastructure—pipelines, harbor work, storage facilities—is already in place. But it will still cost $5.6 billion to convert and will not be operative until 2015, at an initial shipping capacity of 2.2 billion cubic feet/day, roughly equivalent to 117 million barrels of oil a year. Greenfield plants are much more expensive. ExxonMobil is the lead partner on a Papua New Guinea LNG plant: at the 70 percent completion stage in the fall of 2012, the company announced a $3.3 billion cost overrun, bringing the total anticipated cost to $19 billion. In short, these are all high-rolling, high-risk projects.[12]

Economists, by and large, have weighed in almost unanimously in favor of unrestricted exporting, and suggest that it would have little impact on domestic US prices. But virtually every study assumes a high elasticity of gas supply. A study by a Rice University economics institute assumes that every unit of additional gas demand will elicit one and half units of new supply.[13] That is plausible if the bullish projections for recoverable natural gas reserves are true, and they may well be. But only about 10 percent of "technologically recoverable reserves," are actually "proven reserves"—the rest are projections. If it should happen that maintaining supply growth will require tackling more challenging and less accessible fields than we are exploiting now, or environmental challenges succeed in curtailing production significantly, prices could change quickly.

Pressures to approve unlimited LNG exporting built rapidly in 2012. The DOE commissioned an economic analysis of the domestic price impact from a private firm, NERA Economic Consulting, Inc., which was released in December 2012. Wall Street, the industry, and its Congressional minions greeted the report with hosannas, for its overall conclusion, after examining a number of exporting scenarios was that:

The U.S. was projected to gain net economic benefits from allowing LNG exports. Moreover, for every one of the market scenarios examined, net economic benefits increased as the level of LNG exports increased. In particular, scenarios with unlimited exports always had the higher net economic benefits. . . . Natural gas price changes attributable to LNG exports remain in a relatively narrow range across the entire range of scenarios.[14]

Predictably, a small flood of exporting permit requests are arriving at the DOE, and a large Congressional cross-party alliance is developing to sponsor legislation removing any bars to exporting. The report ringingly concludes that exporting will not cause gas prices to be linked to world oil prices. But the *reason* NERA adduces for the low price effects is that if the United States began to export gas, lower-cost suppliers, like Russia, Qatar, and possibly Australia, would immediately undercut American prices and limit its export penetration to quite modest levels. If the industry actually believed *that*, it would be insane to waste untold billions on LNG plants. In fact, judging by trade journal speculations on the vast opportunities on the entire Asian continent, the industry is envisioning a gas export boom.

The industry drumbeat for maximum exports suggests that they really believe they can penetrate Asian and European markets at only a modest discount from current prices in those territories, which would give them a "netback," or revenue after the substantial additional costs of liquefaction, shipping, and de-gasification, that would be far higher than they could earn at home. And if that were the case, they would inevitably raise domestic prices to at least the netback level, which will plausibly be high enough to sink the manufacturing revival. Some members of Congress, like Senator Lisa Murkowski (a Republican from Arkansas) even invoke quasi-moral grounds for exporting, since we "owe it to our allies."

If the NERA report is right, the very precariousness of
the prospects of successful exports is reason enough to hold
up permits. Earning back the cost of an LNG plant takes up
to twenty years. Rushing to build them by the dozen may be
courting the same order of investment calamity as the rush
to build the re-gasification plants. The companies are also
mostly the same ones, trawling financial markets for more
tens of billions in the hope of recovering some of the tens of
billions lost the first time around.[15]

But there is no reason to take the NERA report as dis-
positive on anything, for it is a glib and blindered exercise—
mechanical modeling carried out by robot economists. For
example, the benefits were calculated merely by comparing
the value of new export revenues against the higher gas costs
imposed on the rest of the US economy. Since it was assumed
that companies wouldn't export unless they received higher
returns than at home, that number was necessarily positive.
It simply dismissed out of hand the collateral gains from the
rebound in manufacturing that so many economists anticipate
from the drop in energy prices.

The report is also remarkably tone-deaf. Although it ac-
knowledges that under all export scenarios, "wage income
decreases in all industrial sectors except for the natural gas
sector," it happily concludes that such effects are offset by
higher export revenues and additional income to "consumers
who are owners of liquefaction plants."[16]

Fortunately, as the drive for unrestricted gas exporting was
developing last year, a group of large manufacturing companies, led by Dow Chemical and Nucor, among others, formed
a counterlobby, America's Energy Advantage (AEA), to slow

it down. The economists at Dow Chemical prepared a blistering refutation of the NERA report, which was submitted as part of the record of hearings held on the issue by the DOE in January 2013.[17] Some high points, drawn primarily from the Dow paper, are:

- The NERA report relies on 2009 data from the US Energy Information Agency (EIA), but at almost the same time as the report was released, the EIA virtually doubled its projected rate of growth in domestic gas demand, precisely because of rising manufactures.
- The pipeline requirements for exports do not match up well with those for domestic manufacturers. Just by authorizing the export expansion, the government will trigger an anticipatory shift in pipeline investment that will further disadvantage domestic companies. In such an environment, companies sponsoring new investment would be almost obligated to pull back from their programs.
- The report assumes virtually unlimited natural gas production capacity, which is questionable. Even if the gas is there and readily recoverable, successful legal actions by environmentalists may succeed in greatly lowering actual production.
- The argument that exporting will not drive up gas prices rests on the assumption that the free market will force American competitors, like Russia, to drive their prices down. But gas contracts are usually contracted on a long-term reference price basis, with little scope for market-based adjustments. Beyond that, international hydrocarbon markets have long been cartel-driven, and American oil majors have learned to live comfortably within cartel-dominated pricing regimes.

Indeed, an OPEC-like organization of gas-producing nations
has been organizing to create just such a regime.

- The report dismissed energy-intensive industries as
"low-margin industries," as if they produced little economic
value. In fact, all heavy manufacturing firms do have lower
margins than, say, software and finance. But that's *because* they
have such large employment multipliers. Their margins are
lower because of their heavy supply chain and capital goods
requirements, which reverberate through the entire economy.

- The notion that export profits would be returned to American
"households" is absurd. In the first place, a large number of the
companies participating in the shale plays are foreign—from
China, France, England, Norway, the Netherlands, Australia, and some of unknown origin—and their markups will go
to their own shareholders. In addition, global companies like
Shell and ExxonMobil, although they are headquartered in
the United States, invest globally, wherever they can get the
highest return. There is no guarantee that their foreign revenues will be recycled back to the United States.

In fact, there is a real-life case that suggests that NERA's
and other conventional economic analyses are simply wrong.
Australia has long been involved with natural gas and LNG
exporting, but regulatory responsibility has been delegated
to two different authorities, for western and eastern Australia respectively. The two took different approaches, and so
present a fine natural experiment. The western authorities
designated specific reservations of gas that could not be ex-
ported in order to service local industry at cost-based prices.
The eastern authorities permitted LNG exporting without

any such protections. A detailed study published in October 2012 by the National Institute of Economics and Industry Research, found that, while the western scheme worked more as less as planned, the results in the east were radically different. In the east, as soon as LNG projects neared production, long-term domestic supply contracts "evaporated" as suppliers built stocks for export. Long-time customers were left to compete for surplus gas left over from selling into the east Asian markets. Naturally, the surplus was then available at east Asian prices, the highest in the world, less transport costs. The linkage is quite explicit. Taking into account the decline in competitiveness and employment of the gas-dependent eastern companies, and the negative multiplier effects within the rest of Australia, the Institute calculated that the country loses $24 in economic output for every $1 of gas exports.[18]

In any case, the idea of oilmen extolling the virtues of free-market pricing is oxymoronic. In January 2013, the London Centre for Global Energy Studies (CGES) released a detailed study of the costs of oil production throughout the world. About a third of world oil supply costs less than $10 per barrel and nearly 90 percent costs less than $20 per barrel. On an energy equivalent basis, $20 per barrel is about the same as the cost of producing Marcellus shale gas. So why did the average price of oil hover near the $150 mark for much of the last decade? Partly, no doubt, because of fears of war-related supply disruptions. But the primary reason is that international oil prices are determined by the budget requirements of the oil sheikdoms, with the willing, doubtless delighted, collaboration of their cartel partners like Exxon-Mobil and Shell. For six of the last eight years—every year except for those during the Great Recession—ExxonMobil earned more than $40 billion in profits. As the CGES study suggests, those earnings were generated by pricing power, not cost management.

Let's fill in the rest of the boxes. Why are American natural gas prices so low? Because transport limitations have created an anomalous pricing pocket beyond the umbrella of the cartel. And why are American oil and gas executives pushing so hard for LNG exports, even though the NERA analysis suggests that the opportunity is rather modest? Perhaps the alluring shade under that cartelized pricing umbrella has something to do with it.

The Obama administration has generally been quite cordial to the energy industries. As an old-fashioned liberal who has long lamented the loss of stable blue-collar jobs, I have been glad to see that stance. The tactic has required hewing a middle path between the demands of the environmental and energy lobbies—basically, letting the industry grow while insisting on reasonably attainable environmental standards. Giving the industry its way on this one, however, could completely destroy our best hope in a long time of a broad, middle-class-based economic recovery.

The good news is that there is time. The AEA lobbying, judging by the reactions at the recent DOE hearing on the issue, seems to be having some impact. The CEO of Shell, Peter Voser, said, "Exports will happen. But I hope that the US will actually keep most of the gas back because it will help them to industrialize parts of the US more." Shell is actively planning a large chemical complex in Marcellus shale territory, and has also announced an LNG investment—not for export at this point, but to create a gasoline substitute. (NG transport fuel is practical for fleets with predictable routes and centralized refueling.) Dow has just withdrawn from a partnership in a proposed $6.5 billion LNG terminal. If powerful manufacturing interests continue to join the AEA, broad-based industry groups, like the Chamber and the NAM may well want to modify their unequivocal support for exporting. The AEA also seems to have found the ear of Senator Ron Wyden (a Democrat from Oregon), the incoming chairman of the Senate Energy and Natural Resources Committee, who is holding hearings before taking more definitive steps toward wider exporting. During all this, however, the DOE approved a second Louisiana LNG terminal, for shipping to countries who are covered by FTAs, which legally the Department has no right to turn down. (The company has also filed an application for non-FTA nations.) This second plant is about half the size of the Cheniere facility, and would not be opened until 2018. Recently a senior DOE official told a meeting of state regulators that they intended to "move carefully" on approving such permits.[19]

There is ample justification to put all permitting on hold. The NERA report is hopelessly flawed. It should either be withdrawn or, as Dow suggests, referred to peer review. The United States doesn't have to leap to solve the world's energy problems. Let's look after the home front first, see how well production rebounds as new pipelines open markets and prices rise. If supplies turn out to be as good as the bullish projections, we can gradually let exports run free without damaging the markets at home. In the meantime, Senator Wyden has adroitly suggested approving LNG exporting but only to a level that comports with the forecasts of the NERA report. The "public interest" justification for the limitation would be to avoid wasting tens of billions in building another round of white elephant LNG facilities like the unused de-gasification plants.

But for all the potential economic benefits of the shale revolution, they won't be realized without coming to grips with the industry's dark side.

CHAPTER 3

The Dark Side and How to Deal with It

Perhaps no media event galvanized public unease with the shale gas and oil industry as the release of Josh Fox's 2010 documentary, *Gaslands*, with its striking images of flaming water blooming from rural faucets. The film was nominated for an academy award, and deservedly so. It's beautifully done, and Fox strikes a pitch-perfect balance of sorrow and slow-burning anger, without becoming strident—and plays a terrific banjo besides. Most important, almost all of his accusations are true, although he was fiercely attacked by the industry for his supposed inaccuracies. Some of the fiery water shown in the film most likely acquired its methane from naturally occurring surface sources. (Methane was long known as "swamp gas.") But subsequent peer-reviewed research at those same areas shows conclusively that, while most oil country groundwater will test positive for methane, contamination of groundwater close to shale wells is far worse than in nearby

locations without wells. And methane is among the more be-
nign of the contaminants; the operation of the wells uses large
volumes of noxious chemicals that also migrate into the soil
and surface water.

There are two prongs of the anti-shale argument, however,
and Fox focuses on just one of them—the disturbing side effects of rapid industrialization on water supplies, habitat, the
attendant water and air pollution, plus all the traffic, noise,
and strain on local infrastructure. The second case for the
prosecution, which is quite different, is based on methane's
potentially baleful effects on global warming. I'll deal with
them separately in this chapter, because they are much different diseases.

The Travails of Reindustrialization

The unconventional gas and oil industry is in the mid-stage
of an extraordinarily rapid development cycle. Just a decade or
so ago, the industry was driven by entrepreneurs with limited
capital who were making up the rules as they went along and
racing to generate product as fast as they could. They have
been spectacularly successful in producing new energy at very
attractive rates, but in an unusually intrusive way. The gross
environmental damage caused by the shale industry is far less
than that of coal, and no individual shale well could ever do
the damage that the blowout on the BP Macondo well did
in 2011. But shale forces itself into the public consciousness,
because its product is widely distributed and thinly concentrated. Recovering industrial-scale quantities of shale product

requires venturing far beyond the places where extractive industries typically cluster.

Coal mining and conventional oil drilling usually happen in well-defined regions where daily life is dominated by the extractive activity. When the country singer Loretta Lynn called herself "a coalminer's daughter," she evoked an entire culture defined by the mines. The shale industry model is very different. The region in and around Fort Worth on top of the Barnett shale may be the most intensively developed shale area in the world, with some 16,000 wells within an hour's drive of the Dallas-Fort Worth airport. But the individual wells are small, scattered throughout the area, and easy to miss unless you're looking for them. I recently spent a morning driving around the area, following a rough map supplied by an industry consultant. At first, I didn't see anything at all, then after I picked one out, I realized that they were everywhere, but painted in a kind of camouflage khaki, designed to blend in. These were all mature wells, so the rigs and the engines and trailers, and nighttime floodlights were long since gone. All of the action was subterranean, with the wells soundlessly releasing gas or liquids into underground pipes, possibly throwing off modest amounts of water that is captured in aboveground tanks. Most of them were not much bigger than a decent backyard, just a concrete platform with up to a half dozen Christmas trees—the valve-laden access pipes that rise about four feet above ground—and a small tank, surrounded by a chain link fence. Few of the people living near them would think that those wells are what Fort Worth is "about," and may not think of them at all unless they leak—which does happen, probably more frequently than it should. Several of the wells were backed up right against residential developments, and the people who lived there must have gone through hell during the drilling and completion stages.

During the great days of conventional oil production in the legendary Texas Permian Basin, environmental violations were probably far worse than those of the modern shale industry, but those wells weren't sitting next to somebody's neighborhood. The very nature of the drilling and production cycle offers many opportunities for mishaps—spills, leaks, large vapor releases, loss of fracking water flowback—the kind of problems featured in *Gaslands*. The International Energy Agency (IEA) recently released a paper, "Golden Rules," that suggests rules of behavior that the shale gas industry will need to maintain to retain its "social license" to keep expanding production.[1] The paper's focus is primarily on the "little" problems, like spillages.

But little problems, simple as they seem, can be the most difficult to manage, for they require sustained and consistent management attention, of the kind epitomized by GE's "six sigma" internal quality control system. Managing little problems is also *dull*, and the shale industry isn't good at doing *dull*. Preventing twenty-gallon fluid spills doesn't capture management attention the same way as the discovery of a rich new shale play does.

Shale is a high-speed, high-growth business. The face of the industry, until recently, was Chesapeake Energy's Aubrey McClendon, dubbed by *Forbes* as "America's Most Reckless Billionaire." His company was usually the most active shale driller, drilling roughly 1,500 wells a year with 150 of its own rigs, and participating through various partnerships in as many more, pushing up revenues fifteenfold in a decade. McClendon is an entrepreneurial genius, a plunger and lover of financial leverage, who took his company and its share price on a wild ride. He was finally forced out because of financial arrangements of a kind that are not uncommon in private equity, but raised alarms in a listed company with a half-million shareholders.[2] Although the shale industry now produces more than a third of the country's gas, it still has a wildcatter ethos, a drive for the big hit, a mindset that is ill-adapted to achieving a consistently high levels of execution in the detailed process controls that prevent fluid spills.

Consider a typical mundane but serious problem. The investigative journal *ProPublica* has been running a series on the shale industry for several years now. It recently reported that in the Bakken shale, truck drivers bringing waste fluids to treatment sites frequently dump their loads in vacant fields, especially at night and especially if there is a long line at the treatment location.[3] On our visit to the Devon water treatment center, Sarah Terry-Cobo, the reporter on the trip, raised the question with Robert Brodbeck, the Devon engineer, for she had heard a number of such stories from locals. He pointed out that the drivers using the Devon treatment center were picking up just from Devon wells. So their loads were metered at both the pickup and disposal sites, and the values were paired in an electronic database.

But no other company in the Cana-Woodford exercises that level of control, and the Cana-Woodford, in any case, is the only place where Devon has such a facility. Standard practice is to use contract drivers who get paid by the load. And it's not likely that many companies maintain files of treatment center receipts, least of all the very small companies. A web search for companies operating in the Bakken shale returned seventy firms, all of them production companies, including a large number with only a handful of wells.

This gets at the crux of the industry's challenge. Achieving the energy production levels to support the kind of industrial and economic revival outlined in the last chapter will require sustained development of America's shale energy resources for many decades. But to maintain broad support for the continued expansion of its footprint, the industry will have to get much cleaner, crisper, and more controlled in managing its sites. That will never happen when every new play has seventy, or fifty, or even thirty companies cobbling together little deals for water access or waste treatment with townships up and down a state.

In an ideal world, before any production drilling commenced in a region, the industry would develop a comprehensive water acquisition and recycling plan, together with a staged supporting infrastructure program. The goals would be to make minimum impact on drinking water; to ensure maximum feasible pipeline transport for both clean and contaminated water; and to provide dedicated treatment facilities for all of the industry's fluid wastes. Reaching those goals won't eliminate all spills and contaminations, but it should reduce them to more tolerable levels.

Clearly that kind of organized development process is not going to happen soon, but it's not impossible. Regional development planning could be carried out by a consortium of the companies that own the development rights and contribute fees to fund the upfront costs. Or, conceivably, the industry will come to be dominated by a smaller number of large companies who can internalize the upfront costs themselves. An organized, staged development process shouldn't cost more than the current non-system. In fact it should be cheaper, because it will be more efficient, although it may force the companies to bear some of the costs that are now being palmed off on the local communities.

While the industry has clearly made a lot of progress over the past decade, it still has a long way to go. Public revulsion has come close to crippling the nuclear industry in the United States, and the same thing could happen to shale. That would be sad, for it could scuttle a real opportunity to revive the dynamism of the whole economy.

So the shale industry must deal with a handful of chronic issues, starting with the emotion-laden water question, and including the standard problems that arise at each stage of shale gas and oil extraction, storage, and distribution. For the analysis below I have relied both on the "Golden Rules" paper and on a survey conducted by Resources for the Future, an environmental and energy think tank that found quite a high degree of consensus on the important issues among a carefully selected panel of industry, government, and academic shale experts.[4]

Water Usage and Shale Exploitation

Although it's nearly heresy to say so, pressure on water supplies may be the *least* important problem with the shale energy industry. According to the US Department of Agriculture, farming accounts for 80–90 percent of US "consumptive" water use (water lost to the environment by evaporation, crop transpiration, or incorporation into products). In seventeen western states, farming uses 90 percent or more of all available water, and uses it very inefficiently. About half of all farmland irrigation is performed either with traditional, wasteful gravity methods or with older, inefficient powered methods. In the fall of 2012, Colorado farmers complained that the lack of "a level playing field" was allowing energy companies to pressure farmers' water supplies. A *New York Times* reporter respectfully catalogued their complaints, then noted without comment that irrigation and agriculture consumed 88.5 percent of the state's water, while oil and gas drilling claimed 0.1 percent. The energy companies also paid one hundred times as much for their water as the farmers did.* Even in semiarid Texas, with 100,000 gas wells, the most of any state, the gas industry accounts for only 1 percent of state water usage.

Environmentalists also claim that wastewater from farming and other conventional uses stays within the water supply, while fracking water is consumed forever, since so much of it stays in the ground. The truth is that in most industrial applications, water does not efficiently return to the water supply without careful retention practices, which are uncommon on even the best managed farms. Farm runoff that is not purposefully retained can readily percolate deeply into the soil and become "severely degraded in quality or . . . uneconomic

---

* One the primary reasons for the waste of water in farming is that farmers are undercharged for it. Some arid regions of California are major exporters of rice, which is ridiculous.

to recover." Doubtless, some of it is later regurgitated up as produced water from shale wells.[5]

The entire US energy industry consumes a relatively modest share of the water supply—an estimated 27 percent of nonagricultural water uses—and shale drilling is the least greedy of the thermoelectric processes. Shale gas consumes between 0.6 and 1.8 gallons of water per million British thermal units (MMBTU), a measure of energy produced, while coal mining uses between 1 and 8 gallons per MMBTU, onshore conventional oil wells use between 1 and 62 gallons per MMBTU, and corn-based ethanol consumes up to 1,000 gallons per MMBTU. Nuclear generation is also a large consumer of water, although is not in the class of corn ethanol.[6]

So, why the hue and cry over the shale industry's alleged wasteful use of water? Well, if you're a resident of a small town in Colorado, and it's drought season, and convoys of giant water tanker trucks assemble early every morning outside your window to fill up from the local fire hydrants, you might get upset, even though they had paid for the privilege.[7] It takes 1,000 or more very large tanker trucks for a single fracking job, and when wells are in their active development stage, the trucks can be an overwhelming presence. If the industry could get its act together and work out best-practice regional water solutions, many of these criticisms would fade away.

Construction and Exploitation of Wells

Placement

The first consideration must be the distance of a well from buildings, residential areas, and water sources, especially drinking water sources. Current regulation on this, as on most other aspects of the shale industry, is very spotty.

Of thirty-one states with regulations, as of the fall of 2012, seventeen had specific building setback restrictions, ranging from 100 to 1,000 feet, with an average of 261 feet; three others had setback requirements from specific structures or equipment other than buildings; and the rest had no evidence of regulation. Texas regulations require setbacks of 200 feet.[8] The wells I saw in the Barnett area may have been drilled before the current regulatory framework was in place, because several of them looked a lot closer than that.

Recall that on shale drilling sites everyone, workers and visitors alike must wear fire-resistant coveralls or smocks. Wells have accidents, especially during the drilling and fracking stages. At the time we were visiting Devon, one of their wells in Utah had a blowout that ignited. Some families had to be evacuated, and state troopers closed off an area of a half mile radius.[9] Blowouts are relatively rare events caused by sudden powerful surges from inside the well. All wells use blowout preventers, extra-strong emergency valves at the top of the well. But surges sometimes occur before the valves are in place, and preventers sometimes fail. Characteristically, there are no reliable data on the frequency of blowouts in shale drilling. Offshore drilling is federally controlled, so there is a good blowout database showing that they occur between one and ten times per 10,000 wells when blowout preventers have not yet been set; the different frequencies relate to the stage of the process.[10] Most shale drilling blowouts do not result in fires, but some do. Texas is a state with good data; in 2011, they experienced twenty-one well blowouts, of which three involved fire.[11] Common sense suggests that such operations should be kept well away from populated areas—not out of sight, but at a distance that ensures that civilians are not in danger. A thousand feet, less than a quarter mile, would seem the absolute minimum.

The American Petroleum Institute (API), a research and lobbying group for the industry, produces excellent training manuals and best-practice recommendations for every aspect of the business. Their best-practice suggestion for well setbacks, however, is one that falls short. It is: "when feasible, the wellsite and access road should be located as far as practical from occupied structures and places of assembly." Although I'm sure it's unintended, the infelicitous implication is that you must consider civilian safety only when it's "feasible" to do so.

The second group of setback rules apply to bodies of water and municipal water supplies. Nine states have numerical setback requirements, of which 2,000 feet is the farthest, specifically applying to municipal water supplies. A number of other states regulate setbacks from tanks, holding pits, and the like. The effectiveness of such standards probably depends on local geology. Again, the API has a "feasibility" qualifier on its recommendations, which seems inappropriate.

The mother of all water setback cases, which Josh Fox highlighted in *Gaslands*, is the New York City watershed, which provides high quality, unfiltered drinking water to more than fifteen million people, making it the largest such system in the world. The system was put in place a century ago at great cost, and maintaining its quality has required an assiduous process of land acquisition, zoning, forest management, and other preventive measures, at a sunk cost that must run into the many billions. If the system were polluted, it would take more billions to repair the damage or to provide treatment. Still, the industry is fighting for the right to drill in the watershed, which is just dumb. It shouldn't win every argument, it probably won't win this one, and it needs to develop a better sense for when it shouldn't even try.

Well Construction and Completion

The initial work on a well, which may entail six to nine months of excavating, earthmoving, utility fixing, deliveries of giant equipment, etc., is much like any other heavy construction job—annoying but time-limited. Because of the large volumes of fluids that will be used at sites, however, special attention needs to be paid to water management, including dikes and ditches to control spills and to prevent storm-induced overflows of temporary holding areas for noxious wastes. Site plans normally have to be approved by state regulators.

Once the drilling and completion processes are underway, a host of important everyday issues arise. In approximate order of importance:

- Spillage of greases and diesel oil, and of the noxious chemicals used in fracking (which are more dangerous before they are added to fracking water, since they are much more concentrated). Spillage of noxious flowback and produced water, either from the rig or from holding tanks or pits.
- Leaching of methane, fracking chemicals, or other noxious substances into local groundwater supplies and wells.
- Emissions of methane, by intentional venting in order to manage pressures, or by accidental loss through fugitive emissions.

In modest quantities, methane isn't toxic to humans, but it is highly flammable, and in high concentrations can cause asphyxiation. And because it is odorless and colorless it can readily accumulate to dangerous levels without being detected. More than one house or shed has blown up when methane leached into formations below them and exploded.

- Endangering artesian aquifers through drilling and fracking activities.

Although fracking gets the most publicity—the word is so evocative—the evidence is that far more damage is done by surface spills and poorly constructed wells. For example, the town of Dimock, in the Pennsylvania Marcellus, was one of Josh Fox's prime exhibits. A careful study of several shale gas areas, with a good representation from Dimock, found that 85 percent of the water wells examined in the area contained methane. But the concentrations differed sharply depending on how close a water site was to a shale well, on average by a ratio of 17:1. The *average* concentration of methane in the water supplies in active well areas was already in the range that federal standards recommend for hazard mitigation, while the highest concentrations were well above that.

The high methane concentrations near active well sites immediately suggested contamination by fracking wastes. Its chemical signature was that of thermogenic methane—formed by deep underground cooking rather than by surface bacterial digestion—and was very similar to deep-source methane in the Devonian and Utica shales. But the samples showed no traces of the chemicals and brines associated with fracking. So yes, there was substantial migration of deep-source thermogenic methane into surface waters, but no evidence to link it with fracking. The most likely explanation is that the well casing was leaking.[12]

The only case that seems definitely to implicate fracking in surface water contamination occurred in the Wind River Basin of central Wyoming, dominated by vertical wells—that is, those sunk straight down without directional drilling—owned predominately by Encana. After a series of complaints from local residents, the US Environmental Protection Agency (EPA) performed a study, which involved drilling 1,000-foot test wells, that found methane bound with chemicals that tracked closely with those in the fracking compounds. The study has been fiercely disputed by the industry. A separate study was performed in 2012 by the US Geological Survey, but it was designed to produce additional data for an outside peer review panel to be convened by the EPA, and did not draw any conclusions. The industry and the EPA differ on whether or not those results support the EPA's conclusion. The industry has also attacked the EPA's preparation, preservation, and handling of the samples, and the sample size—the Lance Armstrong defense.

Aside from the chemical matches, the geology of the gas well placement seemed predisposed to fracking contamination. The target shale was only 150 meters or so below the deepest of the municipal wells (it's usually a mile or more below), and there was no intervening cap rock to prevent the upward migration of fluids. Nor did the bottom of the surface casing on the well pipes extend beyond the bottom of the deepest water wells. The quality of the cementing and consistency of the bonding also seems to have been indifferent at best (exhibiting "sporadic bonding over extensive intervals . . .

[including] directly above intervals of hydraulic fracturing.")
With such a configuration, it might have been a surprise if
fracking fluids had *not* migrated into the water wells.[13]

The industry reaction to criticism can be as rote as that of a
consistory of Vatican cardinals. It believes that it is impossible
for fracking fluids to migrate into water supplies, and contests
any challenge to its doctrine to the bitter end. In fact, based
on studies done to date, such contamination is exceedingly
rare, which is a good thing. But it is clearly not *impossible,* although the chemical maelstrom around any energy extraction
area may always raise some ground for doubt.

That circle-the-wagons reflex only damages industry credi-
bility. The energy industry should learn from the airline industry. When there is a serious incident, the airline industry and
its regulators join in convening impartial science-dominated
inquiries. Incident reviews are highly cooperative, and investigative conclusions are highly credible. When the fault for an
untoward incident is laid at the feet of the industry, standard
practices are usually upgraded accordingly. It's the grown-up
way.

Another issue revolving around fracking is disclosure of
the chemicals in the fracking fluid. The industry long op-
posed any such disclosure on "trade secret" grounds. After a
number of states began to insist on full disclosure, the industry organized a web site, FracFocus.org, for voluntary reporting of chemicals. All the major companies are cooperating,
but it is still a work in progress. It is designed to search data
on specific wells, rather than to develop broader data sorts. Its
purpose, according to the industry, is to allow local authorities
and residents to ascertain the chemicals in wells near them,

although the data aren't posted until considerably after well completion. The companies have committed to disclosing only chemicals rated as hazardous by federal or state authorities, and they may still withhold data on proprietary grounds.

In 2011, the DOE's Energy Advisory Board created a Shale Gas Subcommittee to review safety and environmental priorities for the industry, under the chairmanship of John Deutch, a senior professor at MIT who formerly supervised the science and technology branches at the DOE and was later undersecretary of the department. Besides recommending a more readily searchable database, their recommendations came down strongly for "immediate and complete" disclosure of all chemicals used in the fracking process. Many such chemicals have never been subjected to hazard analyses, and so slip past hazardous-only specification, and many "hazardous" rating systems do not capture effects of exposure through environmental pathways. In the Committee's words, the public benefit of immediate and complete disclosures "completely outweighs the restriction on company action, the cost of reporting, and any intellectual property value of proprietary chemicals."[14] It is hard to understand how anyone could argue to the contrary.

Another hot-button issue around well completion and subsequent production phases is the scale of methane emissions. An analysis of methane concentration in an active shale drilling area of Colorado showed a methane concentration in the atmosphere of 3.8 percent, considerably higher than conventional estimates. Many emissions are intentional. Well and pipeline maintenance and repair procedures typically require de-pressurizing, which is usually done by venting—releasing gas into the atmosphere. Similarly, the storage of gas and gas liquids can lead to dangerous vapor buildups that must be vented. Water and gas condensates must be regularly cleared, or "pigged," from gas gathering lines, which requires venting. And so on. There are a variety of tools and procedures for dramatically reducing venting volumes by recapturing and recycling the product, but they do not seem to be in general use. (The EPA argues that they can increase revenue, by conserving saleable product; but the savings are small and are likely sacrificed to speed.) A second issue is whether vented gas should be flared in order to prevent possibly explosive buildups of methane or for environmental reasons, in order to convert methane to CO_2, a less potent greenhouse gas. States differ widely on the topic, while the API recommends flaring all gases that cannot be economically recaptured.[15]

So-called *fugitive* emissions, those that are unintentional, may be far more important. The first recommendation of the IEA "Golden Rules" report is: "Measure, disclose, and engage [with local communities]."[16] I'd like to stress the *measuring and disclosing* part, which bridges the discussion of the immediate impacts of the shale industry and its possible impact on global warming. Over the past few years, there has been a heated academic controversy over the volume of methane emissions from shale gas wells. The kickoff came with the publication of a paper by three Cornell professors that, using the limited data samplings available, projected the life cycle global warming impact from shale gas production. Natural gas has been hailed by many environmentalists as the climate-friendly fossil fuel: its CO_2 emissions are much lower than those of other hydrocarbons, and processed methane is virtually free of particulates, heavy metals, and the other toxic contaminants. Methane, however, is a powerful greenhouse gas, with at least twenty-five times the warming effect of CO_2 (although its lifetime is shorter), and the Cornell paper estimated that methane losses from fugitive emissions were much higher than previously supposed—to the point, in fact, where the global warming impact of shale gas might be worse than that of coal.[17]

I spent most of a day with Tony Ingraffea, one of the paper's authors. He's a professor at the Cornell engineering school, who did his doctoral work in rock mechanics. He is seriously worried about the consequences of runaway climate change in the medium future, and is an advocate of slowing the advance of the shale industry. Unlike many critics, he is both knowledgeable and scrupulous in his use of data. According to Ingraffea, he and his colleagues were shell-shocked at the violence of the reaction to the paper, which began even while the paper was still in peer review. A rushed refutation by one of the federal energy laboratories was trumpeted by the industry and financial press.[18] Many critics attacked what they saw as inadequate data in the Cornell paper, ignoring its conclusion that:

the uncertainty in the magnitude of fugitive emissions is large. Given the importance of methane in global warming, these emissions deserve far greater study than has occurred in the past. We urge both more direct measurements and refined accounting to better quantify lost and unaccounted for gas.

I have read most of the back-and-forth, which has become a footless exercise in competing projections, since there are no reliable data on methane emissions from the gas industry—indeed, there are no reliable data on very much at all, except for basic production and rig statistics. The Environmental Defense Fund (EDF), one of the few environmental organizations that has focused on limiting the damage from fossil fuel production rather than attempting to ban them already has a number of projects under way. Their preliminary research tends to support Ingraffea: while low-CO_2 gas is theoretically an improvement over coal and other fossil fuels from a global warming perspective, their most likely levels of methane emissions wipes out their advantage. EDF has joined with nine major gas companies including Southwestern Energy, Shell, Chevron, the ExxonMobil gas subsidiary, and Encana, among others, along with several universities, to document the life cycle emissions—from well to end use—in a statistically valid way, and come up with best-practice methods to reduce them to environmentally tolerable levels. The first study, to be released in 2013, will establish a baseline for typical emissions at the well production point. The data collection will be conducted throughout the well sites of the nine participating companies, covering a representative sample of wells, geologies, and techniques. A second project, in cooperation with the University of West Virginia, will track emissions from large compressed natural gas (CNG) truck and bus fleets and fueling stations, with results early in 2014. Subsequent studies through the remainder of 2014 will focus on local distribution systems, transmission and storage systems, and gathering and processing facilities.[19]

Most state regulations already require detailed reporting of spillages and other incidents, and the major shale gas and oil states have already formed standards bodies, such as STRONGER (State Review of Oil & Natural Gas Environmental Regulations) to evangelize best practices. Modest amounts of federal support might speed the development of more responsive and publicly available reporting systems than we have now, which would be money well spent.

The shale industry may be at a stage much like that of the personal software industry in the early to mid-1980s. Little companies led the surge of creativity that made the industry possible. Had it not been for the thousands of entrepreneurs who saw the shale opportunity and charged recklessly ahead, it may never have developed in the United States, just as it mostly hasn't anywhere else. The personal software industry sorted itself out within just five or ten years, as companies like Lotus and Microsoft and others imposed a kind of order. The shale industry seems on the threshold of a similar consolidation, probably with ten to fifteen companies controlling almost all the production. As that happens, regulatory pressures should push for area-based development planning rather than just well-by-well permitting, including initial environmental surveys, projected well placement, spacing and well pipe layouts, area-based fluids and waste management programs, dedicated treatment and recycling facilities, comprehensive air and water monitoring devices, and good data collection and dissemination. The sooner that begins to happen, the faster the industry can realize its great promise.

The regulatory apparatus probably should remain centered in the states. Most states with resource-based industry concentrations have strong geologic services, and are intricately involved with setting parameters for mineral rights ownership and protection regimes—to prevent theft of product from a neighbor's plot, for instance. Overarching air and water pollution standards, however, should rest with the federal government, although states could impose higher standards.

The question for the shale industry is not whether it increases air and water pollution, traffic noise, and other disturbances, for it surely does all those things. The question is: What is a reasonable price to pay to reindustrialize the United States? Reindustrialization has been a long-time policy objective of both liberals and conservatives. But industrial jobs are, after all, pretty *industrial.* The golden age of American manufacturing didn't happen amid green meadows. But it was a time of good employment and growing middle-class incomes when smart and hard-working high school kids and community college graduates could get decent jobs and get ahead on talent and hard work alone. Today, even the heaviest industries operate with far less waste and pollution than was true in the 1950s and 1960s. And I would argue that with close attention and committed management, the shale industry could substantially reduce its environmental and aesthetic impact. President Obama has identified a manufacturing revival as a key objective of his second term, and fundamental forces seem indeed to be realigning in America's favor. The energy bonanza may be the critical advantage that makes it inevitable; it would be tragedy to lose the opportunity through simple carelessness in the little stuff.

The second question, the one that Josh Fox only alluded to in *Gaslands,* is: What about global warming?

Living in the Greenhouse

The earth has clearly been experiencing a warming trend for a number of decades now, and there is substantial evidence that discharges from fossil fuels have been an important factor in inducing the warming. Because of the complexity of the interactions of solar activity, ocean tides, the atmosphere, and much else, such things can never be proved beyond doubt. So while it is conceivable that climate change deniers will turn out to be right, the correlation is sufficiently strong that it would be foolish, or thick-headed, to ignore the warming.

So what should we do about it? In part that depends on how the problem is framed. One prominent formulation measures the danger by the volume of CO_2 in the atmosphere relative to the preindustrial age, set by convenience as ending in 1850. At the time, scientists infer, CO_2 loading was 285 parts per million (ppm). Now it is closing in on 450 ppm, which many of the scientific climate-change fraternity believe is close to a tipping point where catastrophic events could begin to happen—like a radical redirection of the Indian monsoons. If that is true, we must start now rapidly reducing carbon emissions all across the globe, for it takes a long time for greenhouse gases to clear. A common calculation is that we have perhaps twenty years to turn the trends around.[20]

But if that science is right, we're doomed; the horse has long since left the barn. China has already surpassed the United States as the world's largest carbon emitter, and India and other emerging market countries are not far behind. There are a couple of billion poor people in the world, who want to have a better life, and a "better life" is generally determined by the amount of energy you have at your disposal: for cooking, heating, delivering clean water, as motive power for machines to do the nasty work, or to allow you to travel, and for bringing you news and entertainment, and classes from MIT. In other words, they want to be like us, and their daily lives often contain horrors that leave little room for worries about climate change.

The scientists and advocates who are pressing for global action tend to speak of what "we" have to do. But the human race is not a "we." Chinese and Indian officials will say all the right things at international climate conferences, but they know that at home they have to keep delivering rapid growth. China's leaders, especially, also know they have to stop the detritus from inefficient coal plants burning low-grade fuel, because it's already killing people, but the focus is on particulates and poisons; any carbon reductions will be a side effect. As emerging countries evolve to the point where services begin to dominate their economies, the energy requirement for each unit of GDP steadily drops. China and India, and much of the rest of Asia, however, may still be a full generation away from reaching that point.

The advanced countries are in no position to complain; it was our pollution that brought things to such a critical point. Moreover, our recent carbon reductions, which appear to be quite substantial, are probably an illusion. If one tracks the carbon content of Western imported goods, as Dieter Helm did for Great Britain, it has risen faster than the advanced nations' carbon reductions. The huge leap in Chinese manufacturing in the 2000s was, from a climate vantage point, simply an episode in rich country outsourcing of industrial pollution. Indeed, given the chaotic state of the Chinese energy sector, the net effect of the Western carbon reductions was almost certainly to increase atmospheric carbon loading. If the United States starts bringing home heavy industries like chemicals and steel, the overall effect may well be to decrease emissions.[21]

Advocates alarmed by the proximity of a tipping point generally recommend extreme actions. Nicholas Stern, an economist who has held senior positions at the World Bank and at the British Treasury, whose report, the *Stern Review*, became something of bible for the climate change movement, laid out a plan whereby all carbon emissions would be reduced to two tons per person per year across the globe by 2050. For the United States, that would require an 80 percent reduction, and for China about a third, since the majority of Chinese still use so little artificial energy. Failure to achieve that, Stern argues, will almost ensure a global catastrophe around 2100. Stern's book *The Global Deal*, laid out a detailed action and funding plan to accomplish that goal, and proposed that the plan be accepted by the "world leaders" and "heads of government" at the 2009 Copenhagen climate change conference, and seems to have sincerely believed that it would be.[22]

Mark Jacobson and Mark Delucchi, an engineer and an economist at Stanford and UC Davis, respectively, have laid out a plan for getting the entire world off fossil fuels within twenty-five years, involving massive development of wind, solar, and tidal power to go along with somewhat expanded hydro and geothermal power.* Their plan contains a lot of interesting engineering—like modeling ways to reduce the problem of the intermittency of wind and solar power by linking regional arrays. They do note that there are "socio-technical impediments" to achieving such a plan, which "may require concerted social and political efforts beyond the traditional sorts of economic incentives outlined here."[23]

An underlying, but explicit, assumption of activists like Stern and Jacobson and Delucchi is that you never discount the future when it comes to humans. If you can avert a trillion-dollar disaster in 2100 by spending a trillion dollars now, you should. But humans, and much less human politicians, don't think that way. We freely undergo hardships for family and children, somewhat lesser ones for our neighbors, and occasionally even for our country, but never for all the people in the world, and particularly not for all the people of the world of ninety years from now. More tellingly, espousing a zero-discount principle for such a distant future requires absolute certainty that you're right. Rightly or wrongly, we have learned to be skeptical of highly certain engineers.[24]

Humans solve overarching fifty-year problems by muddling through, doing what they have to when they finally have to. It is conceivable that that reflex will doom the species to

---

* I admit to being jaundiced about very big, and very tight, engineering schedules. Some forty-five years ago I worked in the New York City government expediting capital projects through the multiple bureaucratic and political hurdles, and helped in various minor ways to advance the badly needed third city water tunnel to the point where it was financed and construction actually started. Construction was supposed to take a decade or so, as I recall. Current expectations are that it will be completed around 2020.

extinction by global warming—an appropriate Darwinian outcome—in which case the earth would have to manage without us. In the meantime, it is surely right that governments should demand environmentally sound behavior, at levels of inconvenience and cost that people are willing to tolerate, while constantly nudging up the bounds of their tolerance. Perversely, the catastrophist literature may just make that harder: if disaster is inevitably on its way—assuming we don't carry out an agenda that looks impossible—then all environmentally friendly measures may seem pointless.

Until quite recently, natural gas was assumed to be a major tool in slowing the pace of climate change, since its CO_2 profile is by far the lowest among the fossil fuels. The Cornell paper has thrown cold water on those prospects—even, I suspect, among its critics. Given the age of our natural gas infrastructure, can anyone really believe that there is not a lot of methane escaping? Upgrading and repairing that infrastructure, installing good measurement and monitoring systems, and ensuring best-practice, methane-conserving methods of depressurizing and controlling vapors, is a big, dull undertaking, which will require constant attention. Yet it may be the best way that the gas industry can make a genuine environmental contribution, by lowering CO_2 with only minor offsets by methane and thereby perhaps win and retain the public good will to allow its continued growth and progress.

Part II

THE REST OF THE ECONOMY

Cycles

The great historian Arthur M. Schlesinger Sr., father of the late historian and presidential adviser in the Kennedy years, may have been the first to identify a cyclical pattern in American politics, one that swung between liberal and conservative poles in roughly twenty-five- to thirty-year cycles. The reason for the cycling, he suggested, was inherent in the imperatives of winning national elections. To put together a national majority, the thoughtful leaders of both the left and right parties—Schlesinger called them "radicals" and "conservatives"—each had to gain the support of a much broader constituency:

> The thinking conservative finds his chief allies in the self-complacency of comfortable mediocrity, in the apathy and stupidity of the toil-worn multitudes, and in the aggressive self-interest of the privileged classes. The honest radical draws much of his support from self-seeking demagogues

and reckless experimenters, from people who want the world changed because they cannot get along in it as it is, from *poseurs* and *dilettanti*, and from malcontents who love disturbance for its own sake.[1]

Political cycles turn when an extended period of either conservative or liberal hegemony brings the baser, more self-interested, or barmiest elements to the fore. The market and regulatory reforms introduced by economic and monetary conservatives in the 1980s, I believe, made a major contribution to the recovery of American competitiveness and economic energy. But as the cycle wore on, conservatism came to be defined by an opportunistic alliance between rapacious finance and militant evangelism. Now the global financial crash has discredited the financial freebooters, while polls suggest that a growing national majority has rejected the right-wing social agenda.

Under normal circumstances, Barak Obama's election victory in 2008 would have initiated a new liberal political cycle that would shift national investment more toward our increasingly impoverished public sector—from bridges to public universities. But policy has been hostage to the large financial deficits stemming from two foreign wars, large tax cuts, revenue losses from the Great Recession, and stimulus spending to mitigate the effects of the crash.

As a solid energy- and manufacturing-based recovery gets underway, however, America's international trade accounts will quickly move toward balance, even as rising incomes and quickening commerce boost federal revenues. Trade and budget deficits will shrink in real terms and cease to dominate the political discourse. A vigorously growing economy going into the 2016 election should lock in a liberal ascendancy for a considerable period. "Liberal" in this context doesn't necessarily mean "Democrat"—Bill Clinton was a distinctly conservative Democratic president in keeping with the spirit of the period. To win national office during the liberal ascendancy, Republicans will make similar adjustments. But true conservatives shouldn't lose heart. It is virtually guaranteed that by the mid-2030s or so, bureaucratization and other ills of the liberal gene pool will have ignited grassroots movements to throw the bums out. But for the next twenty years or so the public sector will be reinvigorated.

In the second part of this book, therefore, I look at two areas in which good public sector policies are critical—infrastructure and health care. Infrastructure almost speaks for itself. The level of infrastructure investment relative to GDP has fallen off dramatically, to the point where it could actually inhibit the industrial recovery.

Health care is a more complex case. It is now one of the country's largest industries and largest single employer, and a field in which, by terms of the Affordable Care Act, the government will necessarily be playing a large role. It is not at all the "stagnant service" or economic "deadweight," that some economists allege; rather it is a technology driver in both semiconductors and biotech, with a high rate of innovation-driven productivity growth—an industry beloved by venture capitalists, a prime target for federal research dollars, and a priority market for major businesses at GE, IBM, Hewlett-Packard, 3M, and hundreds of other companies. And according to *Barron's*, the large pharmaceutical companies may be on the verge of a new round of high-productivity drug innovation, after years in the doldrums after losing patent protection on their last generation of blockbuster drugs. But at the same time, there are deeply engrained cost issues in American health care that must be solved if it is to fully realize its promise.

With reasonable policies, public-sector-driven infrastructure and health care will complement and stimulate the growth in the private sector and ensure that it is more or less balanced across the entire economy. The final chapters offer thoughts on making that happen.

CHAPTER 4

Rebuilding America

The standard Keynesian prescription for an economy suffering from low demand, as the United States now clearly is, is to ratchet up infrastructure spending. The government can borrow at very low rates and build highways and bridges, improve ports, clean up waterways, repair dams, extend commuter railways—in short, undertake a whole raft of public projects that enhance productivity, create jobs, and stimulate spending. Throughout the nineteenth and twentieth centuries, improvements in public infrastructure—from public investment in canals and railways and the opening of silted and tree-clogged western rivers, to the Eisenhower national highway program and the Internet—triggered long episodes of economic growth.

As I argued in the Prologue to Part II, power shifts between public and private sectors are a constant of American history, always in response to specific pressing problems. Success in solving them inevitably builds up an opposite set of issues that eventually trigger a power shift in the opposite direction.

The bipartisan stimulus programs enacted at the end of the Bush administration and the beginning of the Obama administration were clearly successful, if too small to trigger more than a slow recovery. The performance of the US economy, weak as it has been, has been far better than that of the European nations that embraced austerity as the solution to a recession. Ideally, the initial stimulus, which was substantially made up of tax cuts and subsidies for state and local spending in order to generate an immediate income jolt, should have been followed up by a major longer-term infrastructure program, but austerity sentiments within the United States were too strong for that to happen. As we emerge from the recession, however, it is clear that the long period of reduced government activity and very low taxes has left American infrastructure in a bedraggled state. If nothing else, the manufacturing and energy-based revival will require much better roads, bridges, water treatment plants, and other facilities through much of the country.

Chart 4.1 on the next page suggests the extent of the gap. A growing, mobile population, like that of the United States, obviously needs to expand its investment in infrastructure roughly in accord with its inflation-adjusted economic output. As the chart shows, we now spend half as much on public infrastructure relative to the size of the economy as we did fifty years ago, and there is a pressing need to shift priorities.

The most complete assessment of infrastructure deficits is the quadrennial "Report Card" on the nation's infrastructure produced by the American Society of Civil Engineers'

CHART 4.1: Public Infrastructure $ as % GDP, 1956–2007

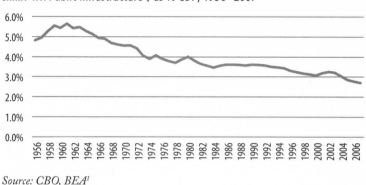

Source: CBO, BEA[1]

(ASCE). It is a serious exercise: a standing committee of senior engineers spends a year working with association staff to organize and update each Report Card, and then they analyze current spending and estimate the shortfall from what is required to bring basic infrastructure to a "good" condition. The 2009 Report Card estimated the total spending requirement over a five-year period as $2.2 trillion, including all current expenditure, which is about $1.1 trillion, suggesting about a $1.1 trillion gap. The 2013 Report Card, released just as this book was going to press, raised the gap estimate to $1.6 trillion. Most of the examples here are drawn from the 2009 report.

═══

The United States has an estimated 85,000 dams, with only about a tenth of them inspected and maintained by the Army Corps of Engineers and other federal authorities. Inspection and maintenance of the rest are at best haphazard. Some states have laws that specifically exempt certain dams from inspection; Alabama doesn't inspect dams at all; and in Texas, each dam inspector is responsible for more than 1,000 dams.

Some 15,000 dams of all kind are classified as "high hazard," a category that is based on the potential consequences of a dam's failure, rather than on the quality and integrity of the dam. Because of population changes and development patterns, the number of high hazard dams has been rising steadily for a long time, while the rate of dam repair has been essentially flat. The consequence is that the number of high hazard dams rated as deficient or dangerous is now well over 2,000; without a substantial increase in the inspection and repair of high hazard dams, that number will grow indefinitely.[2]

━━━ ━━━

Major levees are complex, often very expensive systems that include pumps, drainage systems, tunnels, and other flood-control devices. The failure of the levees in New Orleans during Hurricane Katrina led to damages of more than $16 billion, while failed river levees in the Midwest led to more than a half billion dollars' worth of damage in 2008. The damage suffered by New York City during Hurricane Sandy was probably twice that in New Orleans, not because of levee failure, but because of the lack of serious flood and water surge protections in the first place.

Katrina raised national consciousness of flood protection— as well it might, since 43 percent of the US population lives in a county with a levee—and Sandy has raised this consciousness at least an octave or two higher. Few people in flood country are likely to understand that a "100-year flood" has a 26 percent chance of occurring during the life of a 30-year mortgage. There is no full inventory of the state of the nation's levees; such a degree of inattention suggests that it's not likely to be good.

A partial inventory by the US Army Corps of Engineers suggested that 9 percent of federal levees were likely to fail in a flood event. Many levees are very old, and a large percentage of them were built for crop protection in areas that have long since become densely populated, and therefore need a much higher protection grade. Federal taxpayers have a real interest, since by virtue of a 1968 law the federal government picks up most of the cost of flood insurance payouts. Given the real possibility that we are entering an era of more frequent and more violent storms, serious flood control measures will be much cheaper than paying for more disasters.[3]

The modern environmental movement is sometimes dated from the publication in 1962 of Rachel Carson's *Silent Spring*, which became almost the movement's anthem. The deeper forces are demographic. Over the last half of the twentieth century, the American population grew by two and a half times, per capita water usage increased by 20 percent, and total personal water usage tripled. Maintaining good drinking water and decent rivers and streams became an important national priority.

A wave of federal legislation duly followed throughout the 1970s, under both Democratic and Republican presidents—the Clean Air and National Environmental Policy Acts of 1970 (among other things, creating the Environmental

Protection Agency, or EPA); the Clean Water and Marine Protection, Research, and Sanctuaries Acts of 1972; the Endangered Species Act of 1973, the Safe Drinking Water Act of 1974, a rewrite of water pollution laws that became known as the Clean Water Act of 1977, and the "Superfund" legislation of 1980, which provided for cleanup of toxic waste sites.

There was a wave of new construction throughout the 1970s and 1980s. In the 1960s, New York City had probably the most comprehensive big-city water treatment program, but after his election as New York senator in 1964, Robert F. Kennedy brought pressure on the city to upgrade nearly all of its plants at great expense. Mayors and water authority directors throughout the country presided over a vast buildout of new plants during this time.

Water delivery and waste water treatment systems are long-lived. Underground mains and pipelines have useful lives of up to ninety years, holding tanks and pumps can be in service for up to fifty years, other equipment for thirty years. Once the basic system is in place, especially in locations with a growing population, public pressure tends to bias investments toward extending systems at the expense of maintenance. Systems built in the 1960s and 1970s are aging. At a time of persistent drought through much of the country, public drinking water systems are estimated to leak away up to 20 percent of their product. Funding availability is also much tighter. A signal of how far sentiment has swung came in 2005, when federal legislation exempted the shale gas and oil industry from complying with the Safe Drinking Water Act.

The required investment in drinking water systems and waste water treatment, over and above current expenditure, is estimated by both the EPA, and separately by the nonpartisan Congressional Budget Office (CBO) to be $20–$30 billion a year. Congress created a revolving loan fund to assist state and municipal water system upgrades, but the funds available are only a tenth of the amount required. This is a problem that only gets worse with time.[4]

Much the same story applies to toxic waste sites. The original Superfund legislation that appropriated $1.5 billion annually for site cleanups expired in 1995. Congress continued to contribute to the fund at a somewhat higher level for several more years, but appropriations were cut in the 2000s, and the rate of cleanups completed has dropped with them, from an average of seventy per year in the 1990s to about twenty-five per year in the 2000s, and just twenty-two in 2011. Beyond their health and aesthetic benefits, toxic waste cleanups are good for local economies, since cleaned sites are typically returned to the tax roles.[5]

Rust Bowl manufacturing was heavily dependent on barge transportation on America's inland waterways, a heritage from the heroic nineteenth-century period of rapid Midwestern economic development. When steamboat designers achieved mastery over the region's swift, shallow rivers, they enabled the creation of a tightly integrated riverine farming, manufacturing, and trading economy, built around cities such as Cincinnati, Chicago, Cleveland, Louisville, St. Louis, and Pittsburgh, that created the American pattern of very large-scale, highly mechanized process manufacturing in oil, iron, and steel, as well as grains and meat.

Barge transport is ideal for low value-to-weight commodities like coal and coke, iron ore, crude oil, salt, cement, and other industrial supplies. A single barge can transport the equivalent of 870 truckloads; by a long margin, it is the cheapest form of shipping for non-just-in-time goods. The coming resurgence of manufacturing in the United States will not be an exclusively Midwestern phenomenon, but it will be heavily concentrated in the old industrial states in order to take advantage of inexpensive natural gas. Companies like Nucor and Dow will make heavy use of the inland waterways.

The "system" of inland waterways is a network of navigable rivers and interconnecting industrial canals that, given the long depression of Midwestern heavy manufacturing, has been quietly deteriorating for a generation or more. Major ports include Duluth on Lake Superior, Pittsburgh, St. Louis, Chicago, Detroit, Toledo, and Gary, among others. Interconnecting canals short-cut river connections by using locks to overcome elevation obstacles. About one out of eight locks on today's waterways were built in the nineteenth century, and about half of them are more than sixty years old. Once barges get to their destination, moreover, the efficient movement of goods depends on the local port infrastructure—the wharves, the warehouses, the rail loading yards, the container ports, the highway connections. There is no inventory of such facilities, but all the indicia of age and falling maintenance spending suggests that they are in a sad state.

Compared to drinking water and water treatment facilities, the inland waterways and port facilities are not extraordinarily expensive to restore to fitness. One credible estimate is that somewhat less than $2 billion of additional annual modernization and maintenance spending might be adequate. Currently a fuel excise tax on system users funds about half of all expenditures, with the rest dependent on congressional appropriation. Given the palpable economic benefits of well-directed investment, and the importance of the system to a Midwestern manufacturing recovery, an advisory committee of interested business leaders might be the best forum for working out needs and investment plans financed primarily by user assessments.[6]

The National Highway System (NHS) comprises the interstates, plus almost all (~85 percent) of urban expressways, principal arterials, and other major feeders. It is one of the few infrastructure areas that does not present a totally bleak picture. The NHS comprises only about 4 percent of the total US route mileage, but 44 percent of vehicle miles traveled use the NHS. "Federal-aid-highways" comprise the broader class of roads that are eligible for federal assistance. The quality of the roads has been improving steadily for some time: from 2000 through 2009, for example, vehicle miles traveled on NHS roads rated as "good" rose from 48 percent to 57 percent, nearly a 20 percent increase. NHS bridges have been improving too. "Structurally deficient" bridges—not necessarily dangerous, but requiring significant rehabilitation to stay in service—have dropped from 6 percent to 5.2 percent over that same period, with a comparable drop in bridges classified as "functionally obsolete"—where nothing is structurally wrong, but they are no longer suited to traffic demands.

Traffic congestion has been worsening, however, especially with growth in truck traffic outstripping automobile traffic, and it will be further exacerbated as the manufacturing revival takes shape. Consensus estimates are that merely staying abreast of traffic increases, with minimal congestion improvements, would require an additional $10–$15 billion a year for the foreseeable future. Adding the capacity required to reduce congestion would take much more. In their 2010 "Conditions and Performance" report to the Congress, however, the federal transportation agencies modeled, without recommending, an alternative approach that funded much of the new investment with combinations of congestion pricing and gasoline tax increases. Leaving aside the revenue effects, the investment requirement dropped by 30 percent because of the reduction in vehicle miles traveled, and delivered superior results in reducing congestion delays and the percentage of driving on "rough" surfaces.[7]

Mass transit presents a somewhat different picture. Mileage traveled by mass transit increased by more than 20 percent between 2000 and 2009, and fleet sizes have been expanding at about the same rate. The condition of railcars is considered adequate, and has been stable, while the condition of buses is at the low end of adequate. But there are major deficiencies in the assets that the public does not see, especially in the rail transit sector, relating to switches, signals, power equipment, elevated structures, and the like. Funding for mass transit has been squeezed of late, although it is of great value in reducing congestion. The estimated asset value of the US local mass transit systems is close to $700 billion; about 12 percent of that is past its useful life and should be replaced, in addition to about $40 billion (original cost) substandard non-vehicle assets. Since original costs are a poor guide to actual replacement costs, bringing all plants up to standard would cost in excess of $100 billion, which would necessarily be staged over a number of years. Continuing to build out mass transit to reduce the pressures of local congestion would cost much more. The average commuters in high-congestion cities like Los Angeles and Atlanta spend the equivalent of days each year stuck nonproductively in their cars on a freeway.[8]

Since the deregulation of American freight railroads in 1981, they have been one of America's business success stories—and one of Warren Buffett's favorite industries. The freight rail system comprises some 140,000 miles, virtually all of it managed and maintained by the railroads. Railroads are one of the most efficient of the major transportation industries, moving a ton of freight 469 miles per gallon of fuel consumption. Since deregulation, volumes have roughly doubled, rates per ton-mile have fallen substantially, and profits have been solid. Freight rail employees are among the best-paid of industrial workers: in 2011, they averaged $73,000 a year before benefits; at the same time, for a number of years, the railroads have invested at least $20 billion a year in maintenance and capital spending, and channel 17 percent of their revenues into capital spending, compared to only 3 percent in the average American company.

"Intermodal" shipping—railroad transshipping of containerized truck freight—exemplifies the productivity gains. It is a third the cost per ton-mile of truck freight, faster, safer, and with far fewer noxious emissions. The growth in freight

loadings has raised concerns about the adequacies of the current network. But in contrast to highway traffic, railroads
have considerable ability to improve productivity per mile of
railroad. Just increasing average freight speeds produces an
equivalent increase in capacity, as do longer trains and larger
cars, electronic controls to permit closer train spacing, passing
lanes at critical junctures to obviate slow-freight bottlenecks,
and much more. Smarter railroads may supplant the requirement for more miles of road for the foreseeable future.[9]

The position of the passenger rails could not be more different. While the jury is still out on the economic viability
of long-distance rail travel, corridor services in the 500-mile
range have been quite successful in attracting passengers and
lowering the need for costly and fuel-inefficient short-haul air
travel. The Northeast Corridor between Boston and Washington, D.C., may be the star of the system, but the passenger
rail corridors linking Milwaukee to Chicago and Sacramento
to San Francisco and to San Jose are also heavily used. The
public benefit of shifting high-volume passenger service from
automobiles and short-haul airplanes to railroads, in terms
of greenhouse emissions, fuel savings, and road and airport
congestion, is demonstrable, but passenger receipts will not
support capital improvements.

Passenger railroads have free right of access to roads owned
by freight lines, but they must give the freight lines the right
of way. Amtrak, the national passenger rail corporation, controls less than 700 miles of rail lines, mostly in the Northeast
Corridor. A reasonable public investment program would
aim to provide good quality Amtrak-controlled lines for all
the main corridor lines, with complete separation of through

and commuter train services. Dedicated high-quality rail lines (for example, in concrete beds) would allow the corridor services to cruise at an economically efficient, faster than 100-mph rate. Pursuing "bullet train" technologies, such as magnetic levitation (mag-lev), anywhere other than in completely greenfield sites may be an impossible dream. Just the modest investment program suggested here would carry a total price tag in the $100 billion range and would take a decade or more.[10]

The last major cog in the US transportation system is the airlines and air freight carriers, which are virtually all private entities. Public provision is limited primarily to airports and the federal air traffic control system. Airport financing comes from four primary sources—airport surplus cash flows; tax-exempt bonds issued by local municipalities or airport authorities; passenger facility fees paid by airlines; and government grants. The primary federal grant program is the Airport Improvement Program (AIP) which is sustained by a trust fund supported by excise taxes on tickets, baggage, fuel, and other airport activities. Grants are competitive, and funding is allocated according to priorities established annually by the Federal Aviation Administration (FAA). In fiscal year 2012, $3.3 billion in grants were awarded, most with local matching fund requirements.[11]

The largest federal commitment, however, is its "NextGen" replacement for the antiquated air traffic control (ATC) system. It is an estimated $40 billion project with an implementation schedule that stretches into the 2020s and includes at least twenty separate major initiatives, many of them of daunting complexity. Originally launched in 2004, it quickly foundered, partly because of the refusal of cooperation by the air traffic controllers during a prolonged contract dispute. After the Obama administration settled the dispute in 2010, the program was repurposed and rescheduled and, to the surprise of many, seems to be making solid progress. The goal is to replace radar-based guidance with ground-based satellite links that can track aircraft with far greater precision and allow optimization of flight paths in and out of airports. The prevalent descend-hold-descend pattern would be replaced by much shorter direct descents; flights could be more tightly grouped at both takeoff and landing; and the organization of tarmac sequencing would be much more efficient. If successful, the program should reduce delays, save fuel, and reduce emissions. As of early 2013, practically every critical system has been installed at some airports and in some airlines, to mostly positive reviews from the industry and analysts. The cross-party congressional commitments seem to be such that the program should be relatively immune to the ongoing political and fiscal disputes in Washington. While there are sure to be stumbles ahead as the full panoply of capabilities are rolled out, there are grounds for cautious optimism. Since the vendors are primarily American, a successful implementation should also lead to substantial export contracts as well.[12]

Because the ASCE Report Card focuses on traditional infrastructure, it ignores the sad state of the broadband and mobile communications networks in the United States. A 2010 ranking by Cisco Systems on national Internet quality placed the United States fifteenth, tied with Slovenia. Even Italy and Portugal now have faster and more accessible broadband than the United States. A number of experts believe that the US problem is less related to capital availability than to the preferential position of the (mostly technically brain dead) cable and telephone companies. Much like the old AT&T, without effective competition these companies can sit on their franchises and collect high rents for poor services. If AT&T had not been broken up, and the telephone exchanges put on an open-access basis, the Internet might have been strangled in its cradle. While the public may believe that the Internet is mostly about fast movie downloads, there are good data relating a country's standing as an industrial innovator to the quality of its broadband systems.[13]

The link between economic growth and infrastructure development has also been confirmed in a detailed analysis by economists at the Organization for Economic Co-operation and Development (OECD). Across a wide range of countries, infrastructure investment had a strong relationship to subsequent growth. (The effect was not invariable; the data also turn up negative regressions in specific sectors in specific countries, suggesting likely over-investment.[14])

Moving Ahead

The prospect of wringing very many, if any, new spending programs out of the Republicans in Congress during the second Obama term seems slight, which is a pity. Infrastructure spending financed by borrowing has a long and honorable history in the United States, from Thomas Jefferson, through Henry Clay and Abraham Lincoln, interrupted for a time by

Andrew Jackson's veto of Clay's internal improvement program. Once the Republicans gained power in 1860 and the
Southern contingent departed the government, they quickly
passed the 1862 Pacific Railway Act, which contrary to leg-
end was completed more or less as promised and very close to
the original schedule.

Depending on how fast the recovery takes hold, political
trends suggest that the complexion of the Congress should
shift sufficiently to allow much greater policy latitude, almost
certainly after 2016. Despite the horror stories of "Bridges to
Nowhere," infrastructure investment is usually productivity
enhancing. Better trains, wider roads, faster connections, decent water, clean streams and air—all these improvements pay
for themselves in terms of time saved, fuel not wasted, costs
avoided, and overall quality of life. Standard financial practice
is to borrow for capital projects, amortize the cost, and pay
off the debt over the life of the project. The icing on the cake,
according to economists associated with the New America
Foundation,* is that a program on the scale would create some
5.5 million new jobs. As with new manufacturing jobs, the
multiplier effects of heavy construction tend to be large.[15]

The greatest inefficiencies in public infrastructure investing
stem from "earmarking," the process of aligning budget dis-
tributions to accord with the influence of powerful legislators.
Both parties in the Congress have been attempting to end the
practice, with mixed success. Military base closure commis-
sions have had some success in buffering technical decisions

---

* A wealth of papers on an infrastructure-led recovery is available on its
website, http://www.newamerica.net.

from political influence. Several economists, including Laura Tyson, former chairman of President Clinton's Council of Economic Advisors, have proposed the creation of a National Infrastructure Bank; in Tyson's proposal it would have $25 billion in capital that it could leverage to increase its lending capacity.[16] With a board of directors of unquestioned expertise and integrity, such an institution could play a major role in establishing infrastructure spending priorities, and take the lead in developing innovative public-private partnerships in new infrastructure ventures, especially relating to financing mechanisms for revenue-producing projects.

One fascinating example involves two nearly identical Ohio River bridges now under construction: one is being built as a conventional public procurement, while the other, which originates in a state that allows government projects to be undertaken by public-private partnerships (PPP), is being built by a private entity that will also manage the finished bridge. The same construction company is involved in both bridges and is a partner in the PPP bridge. Predictably, perhaps, it has been much more cost-conscious in the construction and has also recommended and implemented a series of design changes to make the bridge cheaper to maintain. This seems an idea that deserves to spread.[17] (But cautiously; such arrangements were not uncommon in early America. The Pennsylvania Railroad was born from one such successful project, and so was the Union Pacific; however, there were probably just as many PPP projects in which the private partner went bust and left the government to deal with the debacle.)

It will be critical to resurrect the Build America Bonds (BABs) program. They were authorized by the 2008 stimulus legislation, but Republicans have blocked reauthorization since the program entails new federal spending. BABs are taxable bonds issuable by state and local governments to fund infrastructure development. The issuing jurisdiction pays its normal tax-exempt rate, topped up by a 35 percent federal subsidy in order to broaden the market for the bonds. Pension funds, endowments, IRAs, and many other investment entities that are not taxable do not normally invest in tax-exempt paper, but they have been eager buyers of the BABs. The president's 2013 budget includes reauthorization, but with the subsidy falling to 28 percent in two stages.

CHAPTER 5

Cognitive Dissonance

M ost people, when confronted with two perfectly convincing but utterly contradictory interpretations of the world, experience feelings of disorientation and profound discomfort—what psychologists call a cognitive dissonance. That might be a good description of the current American view of the health care industry.

A 2013 *Barron's* investors' report declared health care stocks to be one of "the healthiest sectors" in 2012, with "unstoppable earnings growth" and a "big pickup in the productivity of drug discovery." "Biotech stocks were on fire." Twice as many "very exciting novel compounds" were likely to be approved in 2013, and "a variety of advancements were under way in the treatment of cancer." Indeed, as this book goes to press, a consortium of big private equity funds—KKR, Blackstone, Carlyle—are circling around a possible multi-billion-dollar bid to take over Life Technologies, Inc., one of the world's leading "biotool" producers, working at the intersection of the twin revolutions in microprocessors and biotechnology to produce advanced tools for developing drugs and diagnostics.[1]

At about the same time, William Baumol, a celebrated economist, released a book that characterized health care as a "stagnant service," while a paper by two blue-ribbon health care economists lamented the "deadweight" of health care on the economy. That was consistent with a 2009 analysis by the president's Council of Economic Advisers, warning that the continued growth of health care threatened to drag the entire economy into the mire.[2]

One source of the dissonance is that economists usually count Medicare and Medicaid payments as simple transfers of wealth to the elderly and poor, as the "deadweight" paper seems to have done. But that is not what happens. Suppose an elderly woman gets an open-chest heart valve replacement, which might cost Medicare $100,000. The old lady receives the valve and followup services, but the $100,000 is spread among heart surgeons, operating room nurses, the high-tech firms that equipped the operating room and made the heart valve, and a long list of other medical professionals. The old lady triggers the expense but doesn't make off with the money. Health care is a vibrant industry, with some 14.5 million mostly middle-class workers providing direct services, plus another 1.2 million employed in the pharmaceutical, medical device, and health insurance industries.[3] During the recent, and still-lingering, Great Recession, health care added two million workers. The recession would have been much deeper without them.

One can complain that health care is not efficient and often overcharges, which is true, and I will return to that. But many industries share those faults. Besides being a case study in overcharging, the finance sector occasionally wreaks economic havoc. A London energy think tank reports that the oil companies charge three times the cost of production for their crude, which may be a deadweight record.

The Baumol "stagnant service" hypothesis seems similarly misconceived. Some personal services have little room for productivity improvements. Staging a Molière play takes roughly the same number of person-hours in preparation and performance as it did in the seventeenth century. Kindergarten teachers may also fall into that group, but the concept applies only to segments of health care workers, like home health aides.

Baumol seems to ignore the enormous expansion of health care's technical capabilities. For example, he writes: "people who quit smoking after their first heart attack had much higher rates of long-term survival. Smoking cessation remains a prime means of reducing health-care costs."[4] He has that backward. The data, although sparse, pretty clearly suggest that smoking *saves* health care dollars, not the other way round. Smokers on average die younger and quicker than other people. They are less likely to go on Medicare, and spend fewer years on it. In fact, extending the life of *anyone* who's already had a heart attack, smoker or not, *always* increases health care spending. Saving heart attack patients is one of the major triumphs of post–World War II medicine, and it is fiercely expensive. It is the ever-increasing power of health care technology that is driving the spending increases.

The Expansion of Health Care Services

Anyone who survives a heart attack becomes a permanent resident of the medical mansion. Heart attack survivors need regular prescriptions, frequent specialist checkups, and usually additional interventions, like stenting; many develop kidney problems. They're a quintessential high-cost patient. If extending the lives of heart attack survivors contributed to "reducing health-care costs," private insurance vendors wouldn't search so diligently for pretexts to cut them off their rolls. The range of problems and the methods for attacking them has expanded at the same time—valve replacements (many now implanted without opening the chest), ventricular assist devices for congestive heart failure, genetic therapies for myocardial decline, and within just a decade or so from now, compact mechanical hearts. Productivity in health care doesn't come from washing bedpans faster; it is embedded in the vastly greater powers that technology confers on practitioners.

Oncology is now making a similar rate of progress: survival rates among cancer patients have been rising steadily, and the range of cancers susceptible to successful interventions are growing apace. From 1990 through 2008, the age-adjusted mortality rate dropped by about a third in throat and lung cancers for men, in colorectal cancers in both men and women, and in breast cancer, while prostate cancer death rates dropped by 42 percent. Among the major cancers, the only stable death rate during that period was for women's throat and lung cancers, although the rate was still about a third lower than the improved rate for men.[5] The cost of followup care for cancer survivors is at least as expensive as that for heart attack survivors.

The clearest case of the cost of medical success is AIDS. A few decades ago, the opera diva and impresario Beverly Sills mourned the long string of funerals for talented young artists she had been attending. Once the virus was in your body, it quietly destroyed the immune system, until it triggered full-blown AIDS, which almost invariably led to an early, gruesome death. But over time, an international scientific collaboration, including the major pharmaceutical companies, pushed out the boundaries of bioscience and brought the disease under control. Now AIDS patients who take care of their health and stick to their drug regimens can lead more or less normal lives. But it didn't come cheap, and all AIDS patients are very expensive. Rising healthcare costs are, therefore, partly a measure of the success of pharmaceutical technology. Technological success has created new markets—literally: previously, the customers would have perished.

I got a first-hand look at the state of cardiac technology when I spent most of a year shadowing an elite cardiac unit.* The technological transformation is much deeper than, say, the array of new equipment in an operating room. The assault on heart disease required a cardiac infrastructure far beyond new cardiac ICUs or heart transplant units, including the equipping and training of EMT units and emergency room (ER) staff. A critical ER challenge is knowing *whether* a presenting patient is having a heart attack, for there is a brief

---

* It was in 2006, with the cardiac surgery unit at the Columbia University Medical Center. I observed dozens of operations, went to most of their meetings, including those that dissected case failures, went on a transplant organ run, interviewed patients, and basically had the run of the unit. It is one of the top heart centers in the country, and does the most transplants. Overall, I was deeply impressed with the group's professionalism, their skills, and their ethics. A couple of decades before that, I was part of a venture that developed some of the early for-profit HMOs and so have some appreciation of the cost/quality tradeoffs in any insurance scheme.

window of time after onset when certain drugs are effective. But if the patient is *not* having a heart attack, the same drugs can kill him. Voilà, a bedside test kit to detect the protein marker of a true heart attack. It takes a big and smoothly humming research machine to produce such quotidian marvels. Even the simplest and least expensive therapies aren't cheap. Taking a daily aspirin to forestall a first heart attack has arguably contributed as much as heart surgery to lowering the cardiac death rate. But making that discovery required quite a large standing research capacity, involving both university networks and private companies, capable of managing large population panels over many years.

A substantial slice of academic literature on health care attempts to pin down the productivity of health care by calculating the economic returns from lives saved.[6] With all respect, I think the economists are making a category error—applying a standard from one realm of discourse to another where it isn't appropriate. Leaving aside the simplest cases (e.g., should a professional pitcher be willing to pay for his own Tommy John surgery?), there is no way to measure the economic return of big-ticket health care spending. The "value of a life saved" is just a metaphor; the return, whatever it is, doesn't appear on anyone's income statement. It is also an unrealistic gloss on the real world. The heart surgery patients I saw at Columbia were generally in their late seventies or early eighties, and they mostly did well. The thirty-day mortality rates of such surgeries, even at that age, is only about 2 percent, and about the same percentage end up with major impairments. Some 95 percent or so go home, stiff and sore to be sure, but with their heart problem cured. At the six-week return examination, they usually looked fine, and were glad they'd done it. But they're not going to earn the country hundreds of thousands of dollars in new output; it's more likely to cost much more than that in additional medical care.

We perform heart surgery on young and old alike because we value human life: if we can save a life by a reasonable intervention that is likely to succeed, we do it. (Good people will strongly disagree on the meaning of "reasonable intervention" and "likely to succeed," but the principle still holds.) The shifts in heart, cancer, and AIDS care are high-profile examples, but the same forces have been in play throughout health care, far beyond the treatment of life-threatening diseases. New products and services like lap-band surgery, implantable electrodes to control Parkinson's, injectable drugs to manage chronic psoriasis, all genuinely help people. They also expand the range of medical offerings and increase the customer base. While they have little effect on death rates, they greatly improve the quality of life, but that's what most consumer products do. Refrigerators, cars, computers: medical treatment is another high-tech consumer product that is as desirable as it is marketable. In fact, it's the whole point of economic growth.

The Reconfiguration of American Health Care

The successes of the assault on cardiac disease prompted a number of other big-science "wars" on specific conditions, with coordinated, mostly university-based research, and large infusions of federal money. The rollout of new technologies visibly increased health care spending pressures by the early 1970s.

The Nixon administration responded by pushing health maintenance organizations (HMOs). A cost-saving priority of the
HMO movement was to shift care as much as possible from
expensive hospitals to outpatient centers. "Surgi-centers," or
"Doc-in-a Box" offices started popping up in shopping centers.
The early emphasis was on getting patients out of emergency
rooms to avoid heavy hospital overhead charges. Reimburse-
ment rules were modified to favor ambulatory services—in
effect, splitting the cost-savings with the doctors.

A radical transformation ensued, with utterly unintended
consequences. Vast swaths of medical technology, especially in
surgical practice, were virtually reinvented. Instruments and
procedures were redesigned for minimally invasive practices.
Band-Aid-sized incisions, tiny fiber-optic cameras, tricky
keyhole maneuvering of long-handled instruments are now
completely routine across the surgical spectrum, including inpatient settings. Patients often insist on it—no more big scars.
The movement has gone well beyond surgery; even hospices
are moving away from the inpatient setting. Remote monitoring, along with traveling nurses, perfusionists, and other
professionals, can provide quite a good service for patients
who would rather die in their homes, and usually with higher
provider profit margins.

A 2008 McKinsey research report noted the sweeping
nature of the transformation, which was unique to America.
While the analysts conceded that "the shift has been highly
beneficial" in many respects, they noted with alarm that "due
to its greater convenience to patients and availability of services" utilization was rising rapidly.[7] Health care economists

also frequently lament the lack of savings from simpler, less invasive, surgeries.

But the laws of supply and demand also apply to health care. The old open gall bladder surgery imposed a heavy cost on the patient—a long incision, one or two nights in a hospital, a lot of soreness, and perhaps a week away from work. Most patients had to be doubled over with pain before they decided to do it. With the new procedure, patients are in and out in a half day or so, with only a few Band-Aid–sized incisions, and usually back to work by the third day. So the demand curve shifted. Doctors appropriately advised patients to take care of the problem earlier; there are risks in living with a diseased organ. The data cited by the McKinsey researchers saw no pattern of overpractice. It was simply that the standard of medicine had changed to incorporate a new capability.[8]

That same kind of demand-shifting product development accounts for much of the rise in health care spending. Modern pharmaceuticals are making major inroads in the treatment of depression, hypertension, diabetes, arthritis, and other chronic diseases. Surgical interventions to replace hips, knees, and eye or ear parts, help people stay active and live longer. MRIs have replaced older, invasive, and often dangerous diagnostic procedures, while advanced PET-CT–scanning systems facilitate informed interventions against cancers and heart and brain diseases. In most such instances, if by no means all, product costs go *down*. Treating depression with Prozac and similar antidepressants is cheaper and more convenient than talk therapy, and usually as effective. Since millions of people can take advantage of the new drugs, *spending* on depression has skyrocketed. From the time of Adam Smith, we've known that lower costs and improved convenience expand markets.

But that leaves the question. If technology is improving so fast, why is American health care such a mess?

Free-for-All

Consider two galaxies in collision. The first is the modern American health technology infrastructure, constantly in flux and ever expanding, spinning out one new marvel after the other, some of which will eventually become stars in the medical firmament, but many more that flare brightly for a while then flame out and disappear. That first galaxy is crossing paths with another galaxy altogether, the American medical profession. The collective self-image of American doctors is still redolent of the Norman Rockwell *Post* covers, the solo practitioner, a friend and sagacious counselor, putting in long hours at modest pay, his file drawers stuffed with never-to-be-paid receivables from struggling families. In actuality, however, the medical industry has transformed into a formidable money-seeking organism. Doctors increasingly ply their trade within various forms of corporate organizations devoted to exploiting the hallowed fee-for-service payment paradigm by ping-ponging patients between as many diagnostic tests and minor procedures as they can.

Despite the trend toward group practice, individual doctors are still extremely independent in the details of their practice patterns. While most belong to HMOs and other nominal "managed care" programs, those are rarely more than billing and infrastructure cooperatives. A behavioral research paper by Cornell's Richard Frank reports that doctors tend to develop practice patterns by emulating their peers, rather than from journals or published professional standards, in part because it's the easiest way to minimize second-guessing. He also provides a realistic account of the energy and effort it takes to convert, say, a journal report, into a new practice protocol that runs counter to the local peer consensus.[9]

Since there are thousands, perhaps hundreds of thousands of local physician peer groups in the United States, it is a classic setting for rapid "evolutionary drift." Small variations in initial practice patterns will gradually evolve into something approaching different species of care. Wide variations in practice patterns and costs, in short, would be the *normal* case, not a suspicious anomaly. At the same time, fee-for-service medicine, the American penchant for rapid technical progress in treatments, and pressure from drug and device companies ensure that the drift will generally be in the direction of higher spending.

In real life, when there is a clear and publicized consensus on correct treatment at the medical societies, practice does converge, but it can take a long time. At present, it appears that some 50–60 percent of cases fall squarely into well-defined patterns with only minor practice variations. But that still leaves very large areas where drift is the normal case. There is a similar pattern in hospitals. A recent study using national hospital data on caesarean sections for women who had not had a previous C-section, who were delivering single babies carried to term and not breech, found that the rate of C-sections varied from 7 percent to 70 percent.[10] It has also been extremely difficult to get all hospitals to practice simple, and demonstrably effective, methods of infection control, despite their benefits for patients, for a hospital's reputation, and for the national medical pocketbook.[11]

The surgeon-journalist, Atul Gawande, examined the reality of drift in a 2009 *New Yorker* article, an account of a visit to McAllen, Texas, one of the very highest-spending health care locales in the country.[12] Strikingly, neither the doctors nor the hospital managers he spoke to were aware of their lavish spending patterns. But wherever there is a degree of flexibility in choosing a treatment path, McAllen's doctors seem to pick the most expensive one. The city of El Paso, with an almost identical demographic, has a markedly more conservative practice pattern. There is no lack of venality in the McAllen practice patterns—the doctors themselves cite plenty of examples. But Gawande, correctly I think, pinpoints McAllen as an "outlier."[13] In the evolutionary drift that characterizes US medical practice, *some* locations must end up at the far end of the distribution.

But evolution still has a history. The story Gawande pieced together from his interviews in McAllen was that in the early 1990s a few strong personalities with entrepreneurial instincts became influential in the local societies, and the medical community gradually became more revenue-driven. Over time, they built the most modern facilities, and gradually opted for the most expensive treatments, often with double the national-average volume of procedures and tests. Younger doctors who joined McAllen practice groups took the pattern for granted. Becoming the country's revenue champ was never their objective, but once there, it's an attractive place to be, and they are resistant to change; indeed some engage in ethically ambiguous behavior to preserve the status quo.

Certainly, we should crack down hard on unethical behavior. And there are a number of obvious conflict-of-interest situations, like self-referral, that should never be permitted. One study showed that doctors with ownership stakes in local radiology centers were four times more likely to refer patients for screenings.[14] Promoters of a new spinal insert offering dubious relief to back-pain sufferers pointedly solicit spinal surgeons as investors.[15] Urologists who own biopsy testing facilities order 72 percent more tissue sample tests than a control group.[16] The government has several times declared war on self-referral but has never really seen it through. It would be more appropriate, perhaps, for the profession to adopt it as an ethical canon.

The money-driven behavior of doctors, deplorable as it sometimes is, pales by comparison with the big drug and device vendors and the hospitals that profit by enabling them. Steve Brill, in a splendid, angry, piece in *Time* magazine focused on the bizarre charging schedules that hospitals maintain that incorporate list prices that are five to ten times higher than what they negotiate with Medicare and private insurance companies—for example, $24 for a generic pill that costs a nickel in a drugstore and double-and triple-billing for minor amenities. Aside from extorting large sums from wealthy medical tourists, they weigh tragically on the near-poor who have no health insurance—like the $902,000 bill handed to an uninsured widow whose husband had received home services over the final eleven months leading up to his death. Even religious hospitals supposedly pursuing their mission in "the healing spirit of Jesus" set the lawyers and debt collectors on such people, dunning them for whatever they can get. That kind of exploitation should fade away as the new national

health insurance program takes hold. But Brill's cases highlight the enormous profits built into the supply chain, as well
as the escalating salaries of hospital administrators.[17]

Perhaps the most egregious current example of
hyperexpensive services of questionable benefit is the current
boom in proton beam centers. They cost up to a quarter billion dollars to build, and require a football-field-sized particle
accelerator. Because the US Food and Drug Administration
(FDA) treats most new devices as simply modifications of
the established tools, regulatory approval usually requires
just a showing of the commonalities. So seventeen such centers—costing conservatively $3.5 billion in total—are either
in operation or under construction, and many more are being
planned.

Proton beam centers are an alternative to traditional radiation. Their advantage is that protons are not radioactive, so
while they do penetration damage—like so many tiny, inert,
bullets—they remove the risk of cancer. That makes them
especially attractive for pediatric radiation treatments: since
many radioactivity-induced cancers have long latency periods,
children are more at risk than adults. But pediatric cancer patients are too small a population to justify a nationwide chain
of proton beam centers. The primary proton beam treatment
population, therefore, accounting for about 75 percent of the
patients, are prostate cancer patients, who are predominantly
older men. The assumption—and it is just an assumption—is
that since proton beams are supposed to be more precise than
photon beams, there will be less collateral damage. At least
one proton beam center had to be built to test that hypothesis, but the industry had no interest in waiting for confir-
mation. So far, there has been only one small retrospective

study of prostate cancer outcomes after proton beam treatment. It turns out that they are about the same as with the
standard radiation, and possibly a little worse. A gold standard
double-blind study is underway, but it will be some time before its results are known. The mega-billions boom in proton
beam centers is less about medicine than about market shares
and prestige.[18]

And then there are the drug companies. Besmirching their
reputation is probably not possible at this point; the imper-
sonal villain in current thriller movies is now as likely to be
a drug company as a financial conglomerate. To be fair, big
pharma can turn out wonderful drugs. A Sloan-Kettering
doctor once asked me rhetorically: "Who finally came up
with the AIDS drugs?" The basic research is typically performed in government-funded laboratories, usually in universities. But converting a promising compound into a usable
drug is a science all its own. (Huge swaths of your internal
chemistry exists just to kill off alien invaders like drugs.) So
it is depressing to read each year the list of criminal and civil
settlements the Department of Justice has made with the
drug companies. In 2012, GlaxoSmithKline paid $3 billion
to settle charges that they had unlawfully marketed drugs*

---

* Once a drug is approved by the FDA, doctors may prescribe it for any
purpose, but the manufacturer may advertise and promote it only for the approved application. Marketing drugs for off-label uses is a felony. But companies routinely do it—not by mere hints and suggestions—but in the form of
all-out marketing campaigns with expert spokespeople, in country club settings, with prepared literature, the works. NB: A December 2012 decision by a
New York federal appeals court, found that the marketing rule was a violation
of the marketer's rights of free speech. That could be a dangerous development.
Assuming the ruling is appealed and upheld by the Supreme Court, it could
eviscerate the drug regulatory apparatus.

for unapproved purposes, had misrepresented "best prices" to the government (to get better Medicare prices), and had paid kickbacks to doctors. Abbott Labs paid $1.5 billion to settle similar charges, while Merck paid $441 million to clear up the last charges relating to its Vioxx fiasco. In 2011, the government reached settlements with eleven companies for a total of $1.4 billion, down from $4 billion in settlements against 21 companies in 2010. Rolling the calendar back produces just more of the same—the same companies, the same offenses.[19]

So why do we put up with it? In part because all big businesses have been coddled in recent years, but mostly because Congress won't let Medicare use its purchasing power to negotiate pharmaceutical and medical device prices—which is disgraceful, since it's the norm in all other advanced countries.

So is modern health care just a "Ponzi Game," as an angry senior financial executive said to me a short while ago? No, it's not. To be sure, it's often overpriced and overused. But for the most part it is effective. There have been a number of academic studies of the contribution of technology to improved survival rates among patients. Almost all show strong increases as technology improves—people who have had a stroke, patients with various cancers, neonates, car crash victims. Improved medical interventions have reduced the murder rate by a factor of five since the 1960s. The economist Frank Lichtenberg carried out a series of studies comparing outcomes of like patients by the vintage of drug or other intervention used in the care setting. Patients with newer vintage cardiovascular drugs had significantly lower cardiovascular hospital admissions and lengths of stay and lower cardiovascular death rates. Controlling for demographic variables, a longitudinal study of longevity changes in Medicare patients across all states found significant correlations between longevity and the use of advanced imaging technologies, newer vintage drugs, and the average quality of medical schools attended by physicians. (The centers that used the most modern tools also had roughly the same per capita costs as the ones that didn't.) Two other studies showed that the use of newer technology significantly reduced cancer death rates and improved the "activities of daily living" (ADL) scores of long-term nursing home patients. Finally, in a study of post–age-twenty-five survival rates in thirty countries from 2000 through 2009, the average vintage of pharmaceuticals used was the only variable that significantly related to all measures of longevity improvement; newer drugs accounted for three-fourths of the overall longevity improvement over the period.[20]

In other words, it isn't the usefulness of technology that is the problem, it is the chaotic mismanagement of the overall health care system.

Can We Fix It?

Far more than any other country, the United States has attempted to manage the direction of health care by tweaking payment incentives. It hasn't worked, and unintended consequences, like the vast expansion of outpatient spending and the extreme complexity of billing and payment processes, abound. The worst of the unintended consequences may be that the focus on money has made doctors very conscious of their bottom line. Doctors now talk and think like business school graduates, tracking their market shares and figuring the payback from equipment investments. The whole idea of managing so layered and complex a system by distant manip-ulations of monetary microincentives is misconceived—an-other example of a category error.

Proposals for making health more consumer-driven are equally benighted. Measured by spending intensity, the typi-cal Medicare patient is a little old lady with multiple chronic conditions, on multiple drug regimens that leave her foggy and confused, and possibly festooned with tubing and stitches. There is no central record of which doctors she's seeing or what other medications she's taking. If she is at an end-stage disease point, her family may be insisting on every possible measure to keep her alive, or may be relieved at the prospect of her demise, or may be in violent disagreement with each other. In such a setting, "consumer-directed care" is a nonsense.

The clear pathway to reforming American medicine is to replicate true managed care plans along the lines of Kai-ser Permanente and Mayo Clinic models. Doctors work on salary, so there is no jockeying to maximize fees. At Kaiser,* there is a conscious injection of an overhead medical man-agement layer—run by doctors, not economists or B-school grads. Senior Kaiser doctors in each specialty regularly assess best-practice standards, and over time have developed dozens of "care pathways" covering the bulk of presenting patients. Innovative treatments are adopted when there is evidence to support their use; staff doctors may not adopt any hot new

---

* I have spent time researching the Kaiser practices, but have never visited Mayo, so I don't know the details of the Mayo protocol management process; although from everything I have heard they operate in a broadly similar way.

idea. The care path also stretches across the institutional spectrum, from outpatient through inpatient and rehabilitative care. In other words, unlike the standard American group practice—solo practitioners sharing common housing and office facilities—Kaiser is a real *system*. Patients are not left to find their way through a medical maze; instead, they are directed by Kaiser, and are monitored through each stage of the care pathway.

Kaiser also tracks outcomes by doctors and treatment modality to feed back into the care pathways. Peer review meetings keep medical performance under continuous surveillance. There's no black magic here. The heart surgeons at Columbia do that too, but they have a small, contained, operation. The Kaiser achievement is to bring it to scale. Kaiser plans are not necessarily the cheapest care; they tend to be roughly in the middle, which sounds right if they are reducing excessive care while using case tracking to ensure care follow-through.

Now that Obamacare seems to be secured, the best path for reform would be to gradually convert the state exchanges, and all of the federal exchanges, into Kaiser/Mayo-style staff-doctor practice plans. A number of surveys suggest that doctors are becoming more willing to work on a salaried basis; the gradual feminization of the profession also pushes in that direction, since it's easier to combine professional and family lives. Over a longer time frame, assuming that Obamacare evolves in such a direction, it would make no sense to maintain Medicare and Medicaid separately. As it stands, a large number of working poor are likely to bounce back and forth between Medicaid and Obamacare. Their lives are hard enough without such complications. The same is true for seniors. Kaiser/Mayo-type programs may want to start geriatric planning for some people before the age of 65, but there may be little point in doing so if the patient is changing systems at her next birthday.

The administration's "accountable care organizations," (ACOs) may possibly be a way-station toward a true group practice, but they are unlikely to be successful as they stand, since they are another version of the strategy of managing fee-for-service doctors with overriding monetary incentives, which has never worked. But they might turn out to be a useful transitional platform. If practice groups with strong internal links gain a major role as exchange service providers, it may be easier to take the final step and reorganize them as true managed-care programs.

My guess is that some 60 or 70 percent of American patients will gradually be drawn into federally-sponsored managed care plans, which provide evidence-based standard care at reasonable prices. The result will be a distinct tiering of care plans: one for the average joe, and a large array targeting the moderately well-to-do through the upper 1 percent. There's nothing wrong with that. The government shouldn't subsidize luxury care, and in particular, should not allow company-supplied luxury care to be tax deductible. (Or the level of the deduction could be limited to the amount of the average cost of the standard federal programs.)

Such an evolution, however, would require significantly upgraded data systems to track and monitor care pathways. The Kaiser system has been evolving for some twenty years now. It ought to be embarrassing that the United States, the world leader in applying information technology to business challenges, is dead last among the industrial countries in applying systems to health care. A second major benefit of good data systems is that newer techniques of statistical analysis can simulate long-term, large-population studies of procedures, devices, and medications that are not feasible with the current reliance on clinical trials.* (It was the Kaiser data systems that first flagged the problem with Vioxx.[21])

To get the most out of such systems, Democrats would have to drop their opposition to medical malpractice reform. Standardized case records, with only patient names expunged should be available to public authorities for quality analysis and empirical quality ratings, just as airline flight records are. Compensation for injury would be best administered through a workmen's compensation-type system with clear and consistent standards for awards.

There are other straightforward cost savings that could be implemented with the stroke of a pen. One would be to sharply restrict "Medigap" insurance, that covers all or most of the Medicare copays and deductibles. It has been amply documented that it greatly increases utilization. The impact

---

* The modern "gold standard" for clinical effectiveness is the randomized, double-blind clinical trial. In real life, it is a highly porous, easily finessed, hurdle. Drug companies are masters at manipulating trial size to disclose effectiveness against a placebo while missing remote side effects. Large-sample, long-term, trials are expensive and difficult to administer, and trial subjects tend to be younger and healthier than typical patients. The most powerful of the newer statistical techniques may be "propensity scoring," which facilitates the retrospective creation of control and trial groups from large patient databases. While the comparisons are not as clean as those from a true double-blind study, they can be built relatively cheaply, can quickly analyze very large populations over much longer time periods, and are usually far more revealing than the trials used for FDA approvals.

on low-income seniors could be mitigated by means-tested copay and deductible schedules.

There are, at the end of the day, three reasons why health care in America costs more than in other countries. Medical professionals generally get paid more in the United States than in other countries, but so do lawyers, veterinarians, and most other professionals. That won't change. Secondly, the radical shift to outpatient service provision brought more convenience, less trauma, and usually lower prices. That is a good thing. Sales naturally went up, which is not awful either. Other countries are gradually moving in the same direction, and their costs are now mostly rising faster than in the United States. (In an analysis of per capita health spending in twenty-six OECD countries between 2004 and 2010, twenty-one of the twenty-six had higher growth rates than in the United States.[22]) Finally, the freewheeling process of American technology adoption generates a lot of wasteful spending on untested procedures that are highly profitable for drug and device manufacturers, undermine the ethics and integrity of the professions, and often enough put patients at risk.

It is this last feature that should be the object of public policy. And we know how to fix it. Make better use of federal bargaining power in setting vendor payments. Establish a more organized context for medical practice, preferably along the lines of the Kaiser/Mayo group practice models. That will take a while, but the process could be expedited by consistent pressure in that direction through the new Obamacare machinery.

CHAPTER 6

WrapUp

The consensus prognostication for the American economy is still pretty grim. The Congressional Budget Office (CBO) thinks that real growth will climb back to the 3.0–3.5 percent range in 2014–2015, which is still slow for a recovery, but then will slip back into the same 2.2–2.5 percent doldrums that we're in now.[1] The most recent forecast from the Organization for Economic Co-operation and Development (OECD) for annual average real US growth for the period from 2012 through 2030 is a similarly dismal 2.45 percent.[2] Those are depressing numbers, but they're not unreasonable if you accept the underlying growth narratives. Their core assumption is that the economy will struggle along more or less in its current mode, but with a shrinking labor force as the boomers retire—a long twilit road to nowhere.

But there is now a very different and much more compelling growth narrative. It has four main elements: the energy bonanza; the resurgence of manufacturing; an infrastructure build; and a vibrant health care industry—more organized and with the abuses cleaned up, but still growing. All four come brimming with promise of creating well-paying middle-class jobs. I don't worry about our "shrinking labor force." Male labor force participation rates, at all ages, have dropped sharply in recent decades, but that should reverse itself with a broad expansion of decent-paying jobs in traditionally male occupations.

The second two, infrastructure and health care, are much more government driven. At the moment, both are being squeezed by deficits, which by consensus forecast will hang around our necks forever. But growth is a magic cure for deficits: with growth in place the government can continue to invest, and infrastructure and health care are both areas in which government spending is not "deadweight"—it goes right back into the private economy. The economist Uwe Reinhardt, who has made a specialty of health care, points out that:

> At the local level, policymakers usually give much weight to the employment opportunities offered by a growing health sector, which leads them to resist reductions in or closing of local health care facilities. On the other hand, at the macroeconomic level, policymakers often view growing health spending with alarm, although added consumer spending on other goods and services—on SUVs or entertainment—invariably is viewed as a sign of economic health by both policymakers and the media.[3]

The infrastructure investments, in particular, will be essential if we are to realize all the potential benefits from the energy boom and the manufacturing renaissance. We need upgraded transmission lines, better grids, fewer transportation bottlenecks, modern bridges, and sufficient waste treatment capacity. Infrastructure investment has always been a key contributor to periods of rapid US growth.

Fumbling the Manufacturing Future

How could we screw up the prospects for energy and manufacturing? Getting energy right is paramount. While the manufacturing recovery will happen independently of the shift to native low-cost energy, the energy bonus makes the American advantage overwhelming. There is a vast literature on the emergence of industrial clusters—the regional networks of suppliers, consultants, school-based training programs, tailored infrastructure, knowledgeable shippers and other middlemen, and trade organizations that disseminate best practices. As clusters deepen, they attract more companies and increase the locational attachment of the ones already there. Network germination is always partly spontaneous, but usually there is some central attraction. In the early days of Silicon Valley, the willingness of Stanford University to assist entrepreneurs with space, graduate students, and other support was important. In the German Mittelstand—the middle-sized, export-oriented manufacturers—the local educational systems target apprenticeship and other training. But once critical mass is reached, networks become self-subsisting, and clusters continue to expand on their own. The availability of cheap energy is attracting a long list of early movers to the United States, perhaps already in numbers sufficient to generate the kind of self-sustaining network effects that ensure long-term advantage.

It will also be essential for the oil and gas companies to develop a détente with the environmental movement. Not every wisp of gas or drop of oil will be accessible. Consolidated, efficient water transport and disposal will be necessary to reduce surface spills. Companies must be much more forthcoming with their data. The alliance between the Environmental Defense Fund and nine major exploration and drilling companies to document life cycle methane emissions bodes well for the future. But if the tests show large-scale emissions, as I suspect they will, they will have to respond with substantial upgrades of their plant and procedures; otherwise, exploration could be stopped in its tracks.

But the worst threat is the drive toward massive exporting of liquefied natural gas (LNG). Deep in its corporate heart, the industry evidently desires to become the chosen energy supplier to all of East and Southeast Asia, reaping $17/MMcf prices, with a $10–$12 net return (equivalent to selling oil at about $75 per barrel). The LNG applications already waiting in the queue would absorb at least half of potential American production. At that volume, local and foreign prices would assuredly link at about the net return level, and America's energy advantage will have been thrown away for the sake of global energy company profits. In unguarded moments, that's what canny oil men say will happen.[4] That's also why the companies have all shifted to wet gas and natural gas liquids (NGLs) production. NG-based propane is still propane, and they sell it at petroleum-derived propane prices and pick up three or four times the margin.

As this book goes to press, a new, highly granular paper by Charles River Associates not only provides a devastating analysis of the 2012 NERA report, which the industry is relying upon to make its case, but virtually accuses NERA of cooking its outcome, by making unwarranted assumptions that "strong-armed the model into producing [the favorable] results."[5]

If the companies succeed in realizing their LNG dreams, the United States could wind up as a raw material colony of an Asian industrial juggernaut. Great Britain once foresaw such a role for its North American colonies, and we fought a war to prevent it. It would a sad day if the oil majors succeeded where Great Britain failed.

Financing Health Care and Infrastructure

The major obstacle to expanded investment in infrastructure and health care is the state of the American balance sheet. The two Bush tax cuts, two wars, and the intentionally underfunded Medicare Part D program left yawning deficits that were only worsened with the collapse of revenues after the financial crash. Stimulus spending cushioned the worst effects of the crash, but at the price of even bigger deficits. President Obama was elected on a progressive platform of infrastructure investment and health care reform, but his hands have been tied by the deficits.

Then again, even the biggest deficits are susceptible to growth. Chart 6.1 shows a real-life illustration of the magic power of growth. The highest debt-to-GDP ratio in American history was in 1946, when wartime borrowing pushed the

CHART 6.1 GDP and US Debt: 1941–1956 (current $)

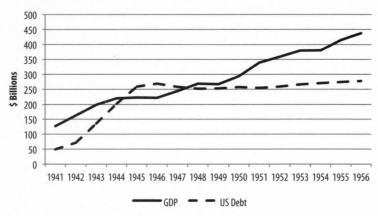

Source: *Bureau of Economic Analysis (BEA); US Treasury*

value of outstanding debt to 121 percent of GDP. But when growth took off after the war, it quickly ceased to be important.

The position today is not much different. The deficit, while not nearly as high as in World War II, has risen to a point where it is worrisome enough to limit the country's ability to invest in needed infrastructure or to offer decent social safety net programs. The CBO now sees this as a more or less permanent condition. But let's assume just 1 percent additional real growth, which takes us close to 3.3 percent steady-state growth rate since 1950. (Over the next decade, at least, we should do better than that.) The charts below start with the CBO assumptions on spending, inflation, and taxes, and adjust their growth assumptions by just 1 percent, and the difficulties disappear. (In Chart 6.2, the GDP forecasts are the upper pair.)

CHART 6.2 CBO and Adjusted GDP and Debt

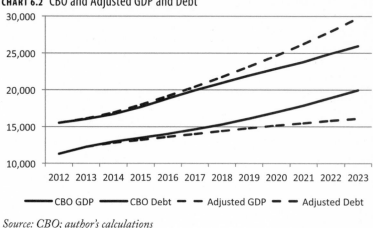

Source: CBO; author's calculations

And Chart 6.3 shows what happens to the debt:

CHART 6.3 CBO and Adjusted Debt % GDP

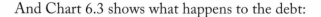

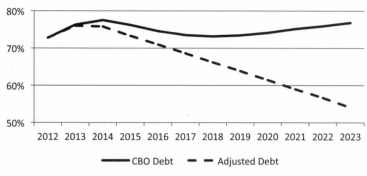

Source: BEA; US Treasury; CBO; author's calculations

Federal debt quickly flattens out in real terms, and steadily falls as a share of GDP.

Health Care and Crowding Out

The growth in health care spending in recent years has also raised the specter of "crowding out." In theory, if health care's share of GDP keeps rising it will squeeze out other more useful activities. But the share of income devoted to specific sectors is always changing. There was a time when Americans spent half of their income on food. Chart 6.4 shows the share of personal income devoted to "necessities"—housing, food, clothing, and health care—from 1930 through 2010. Since 1950, inflation-adjusted per capita income in the United States has increased by 3.3 times, and within the four necessities, health care and food have switched places, not because people are scrimping on food, but because agricultural productivity has made food so cheap. Clothing has made a similar, if not so pronounced, productivity-based reduction in share. The share of income available for "all other" has grown throughout the period. (Note that when incomes *fell* after the crash, spending for "all other" lost share, which one would expect. The critical thing is to keep incomes rising.)

In a 2009 paper, the Council of Economic Advisers specifically raised the fear of squeezing out. They forecast that health care spending might rise to 40 percent of GDP by 2040. (So far it has risen more slowly than they expected.) If we accept that forecast and assume that the economy rises at its historical post–1950 real rate of 3.3 percent, what are the consequences? Health care spending nearly quintuples over the three decades, but resources for everything else still double. And the "everything else" will include some dirt cheap things that are now quite expensive, like high-quality portable

CHART 6.4. Share of "Necessities" in US Personal Spending, 1930–2010

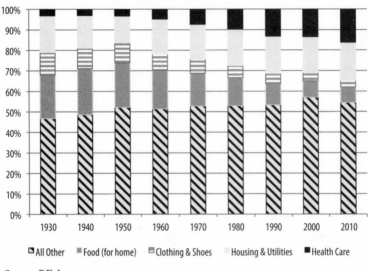

☑ All Other   ▤ Food (for home)   ▤ Clothing & Shoes   Housing & Utilities   ■ Health Care

Source: BEA

entertainment devices, and some very expensive new things that we've not yet heard of. If the economy slips into the doldrums that the CBO and the OECD expect, however, then the thesis of this book will have been shown to be wrong, and spending on everything, including on health care, will be curtailed.

Regardless of its economic effects, it may still be necessary to limit federal spending growth, almost out of principle. But there are plenty of targets on the government accounts. During the Bush years, real annual spending on the military increased by 76 percent, or 3.5 times faster than the economy.[6] The 2009 military budget inherited by President Obama was $300 billion bigger in real terms than the one President Bush inherited from President Clinton. Military budgeters tend to treat each new budgeting high point as an untouchable minimum. US military spending is now so high—roughly equal to that of the entire rest of the world—that it is ripe for major trimming.[7] For example, few people recall what a modest military presence we used to keep in the Middle East. The real buildup didn't start until the 1991 war in Kuwait, and has burgeoned mightily since then.

The withdrawals from Iraq and Afghanistan should trigger a major downsizing of the forces in that theater. Unlike Great Britain, we never (allegedly) wanted to become an imperial power, so we may wish to gradually back away from our role as world's policeman. Safeguarding the petroleum sea lanes benefits the entire world—in fact, however, one could argue that it's not nearly as important to us as it once was. There are a lot of other rich countries, and it's time for some sharing.

Taxes and Deficits

Few Americans realize how lightly we are taxed. Economists at the OECD regularly tabulate a league table of major country tax impositions. The calculation involves *all* taxes, at *all* levels of government. And the rates are calculated as a simple percentage—the actual amount of all taxes collected divided by the country's GDP. Chart 6.5 displays the most recent table.[8]

There is a wealth of evidence demonstrating that non-extreme changes in tax rates have little measurable effect on output. As the chart shows, a lot of successful countries get along with taxes schedules that are much higher than ours.

CHART 6.5

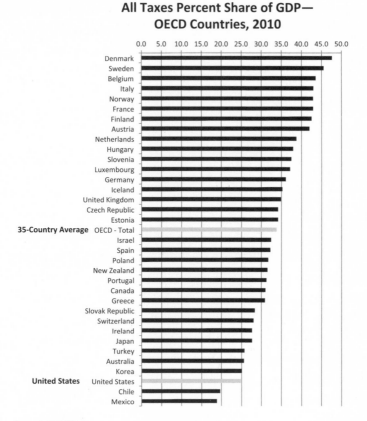

All Taxes Percent Share of GDP— OECD Countries, 2010

Source: OECD

The next chart shows the divergence between US and other OECD tax levels over time.

The widening US/OECD tax divergence during the 2000s is primarily because of the large tax cuts in the 2000s, initially at the federal level and then in many states. At the same time, spending, primarily for the military and a new Medicare drug benefit, went up sharply. It seems clear that the tax cuts were

CHART 6.6

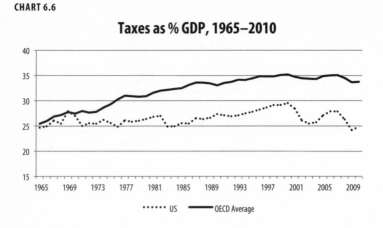

Taxes as % GDP, 1965–2010

inopportune. It is moreover profoundly undemocratic to reduce taxes as one increases spending, since it blocks policy
cost signals back to citizens. In any case, there is clearly room
to move the rates back up a good notch to maintain vital infrastructure investment and services like health care.

Finally, a truly balanced budget is not an appropriate objective in the modern era. *Some* level of deficits, especially in
the United States, is probably essential for well-functioning
financial markets. At the end of the Clinton administration,
as surpluses began to build up, there was consternation in the
global financial industry as American treasuries got scarcer,
for they are the standard risk-free reference for all complex
investment portfolios. Then-Federal Reserve chairman Alan
Greenspan argued the necessity for the Bush tax cuts in order
to maintain the global supply of treasuries.

As a matter of public policy, I believe major public programs, such as Medicare and infrastructure investment, should
be financed from dedicated taxes, so people have a personal

idea of what they cost. If Congress were ever to undertake a thorough tax overhaul, this is what I would recommend.

Social Security should continue to be financed from the payroll tax. Some modest changes, like increasing the payroll tax ceiling and changing the bend points for the better-off, would be sufficient to get it back into actuarial balance. The Bowles-Simpson deficit reduction proposal is excellent.

Medicare, Medicaid, and all other government health care subsidies should be fully funded in a separate trust fund financed by a consumption tax. Economists have long pointed out the virtues of a consumption tax, and health care is a major consumption item. Consumption taxes need not be regressive; they can be made as progressive as we please, by exempting lower incomes and phasing rates as one moves up the income ladder. They should also be simple to administer.

Infrastructure spending should be financed by a dedicated carbon tax—in effect, an expansion of the current gasoline tax, covering all hydrocarbon-derived energy, ideally on a value added tax (VAT) basis. The proceeds should cover all federal non-military infrastructure spending and research, including energy research (but not business development, which is often not a good idea anyway).

Both the consumption and infrastructure taxes should be calculated to fully cover each year's planned spending, including debt service on capital programs, plus the deficit from the previous year.

The income tax could then be much reduced and still leave room for a greatly expanded national research effort in computers, biotech, and other fundamental sciences. One could do much worse than to simply reinstate the 1986 Reagan tax bill, which eliminated great swaths of high-end tax preferences, with the appropriate adjustment in tax rates.

———————

The United States is on the threshold of a long-term economic boom, one that could rival the 1950s–1960s era of industrial dominance. Dominance was easy in that era; when it started, the United States stood alone in its wealth and industrial supremacy. Those days are long past, and the world teems with rivals. But the combination of rising American productivity and the new X-factor, the American energy advantage, offer huge opportunities. We can rebuild our middle classes, provide reasonable social safety nets and health care, shore up our sagging infrastructure, and get our national debt under control. That's what economic growth can do.

But we can mess it up. The opportunity is clear; the pitfalls are known. It's up to us.

Notes

PART I: PROLOGUE

1. Company financial reports and investor presentations.

CHAPTER I

1. International Energy Agency (IEA), *World Energy Outlook: 2012* (Paris, France, 2012), 81.136.

2. Alex Trembath et al., "Where The Shale Gas Revolution Came From: Government's Role in the Development of Hydraulic Fracturing in Shale," *Breakthrough Institute Energy & Climate Program* (May 2012), http://thebreak through.org/blog/Where_the_Shale_Gas_Revolution_Came_From.pdf. For a taste of the government's massive role in the development of coalbed methane technology, see "Proceedings of the 2nd Annual Methane Recovery from Coalbeds Symposium," Morgantown Energy Technology Center (MERC), April 1979, http://www.netl.doe.gov/kmd/cds/disk7/disk2/MRC%5CProceedings %20of%20the%20Second%20Annual%20Methane%20Recovery%20from%20 Coalb.pdf.

3. For an introduction to the technology of shale gas drilling, see Charles Boyer et al., "Producing Gas from Its Source," *Oilfield Review* (Autumn 2006): 36–49.

4. D. J. Mossman, Bartholomew Nagy, and D. W. Davis, "Comparative Molecular, Elemental, and U-Pb Isotopic Composition of Stratiform and Dispersed-Globular Matter in the Paleoproterozoic Uraniferous Metasediments, Elliot Lake, Canada," *Energy Sources* 15:2 (1993): 377–387.

5. Jeffrey S. Dukes, "Burning Buried Sunshine: Human Consumption of Ancient Solar Energy," *Climatic Change* 61:1 (2003): 31–34; US Energy Information Administration (EIA), "Unconventional Natural Gas Liquids" in *Annual Energy Outlook* 2006 (AEO2006), (Washington, DC, 2006).

6. Rick Lewis et al., "New Evaluation Techniques for Gas Shale Reservoirs" (paper presented at Schlumberger Reservoir Symposium, 2004).

7. Except as noted, shale descriptions from EIA, "Review of Emerging Resources: U.S. Shale Gas and Shale Oil Plays" (Washington DC, July 2011).

8. United States Geologic Survey (USGS), "Assessment of Undiscovered Oil and Gas Resources of the Ordovician Utica Shale of the Appalachian Basin Province, 2012," National Assessment of Oil and Gas (September 2012), http://pubs.usgs.gov/fs/2012/3116/FS12–3116.pdf.

9. USGS, "Assessment of Undiscovered Natural Gas Resources of the Arkoma Basin Province and Geologically Related Areas," National Oil and Gas Assessment Project (June 2010), http://pubs.usgs.gov/fs/2010/3043/pdf /FS10–3043.pdf.

10. Rex W. Tillerson, Chairman and CEO, Exxon Mobil Corporation, "The New North American Energy Paradigm: Reshaping the Future" (on-the-record presentation at the Council on Foreign Relations, June 27, 2012).

11. EIA, *Annual Energy Outlook 2012* (June 2012). See discussion "11. U.S. crude oil and natural gas resource uncertainty" *ff.* 56–59; USGS, "Assessment of Undiscovered Oil and Gas Resources of the Devonian Marcellus Shale of the Appalachian Basin Province, 2011," National Oil and Gas Assessment Project (August 2011), http://pubs.usgs.gov/fs/2011/3092/pdf/fs2011–3092.pdf.

12. Rafael Sandrea, "Evaluating Production Potential of Mature US Oil, Gas Shale Plays," *Oil & Gas Journal* (December 3, 2012).

13. "New Rigorous Assessment of Shale Gas Reserves Forecasts Reliable Supply from Barnett Shale Through 2030," University of Texas Press Release, http://www.google.com/#q=rigorous+assessment+barnett&hl=en&source =univ&tbm=nws&tbo=u&sa=X&ei=gCQ1UdWkGoXN0wGpiYDgCQ&sqi =2&ved=0CC0QqAI&bav=on.2,or.r_gc.r_pw.r_qf.&bvm=bv.43148975,d.dmg &fp=cae25ae9f45ad1e8&biw=1218&bih=920.

14. IEA, *Golden Rules for a Golden Age of Gas: World Energy Outlook: Special Report on Unconventional Gas* (Paris, OECD/IEA, 2012), 28–30; Hemant Kumar and Jonathan P. Mathews, "An Overview of Current Coalbed Methane Extraction Technologies," Department of Energy & Engineering, Pennsylvania State University (2010), http://www.netl.doe.gov/kmd/RPSEA_Project _Outreach/07122–27_CBM_Initial_review.pdf.

15. Alberta Chamber of Resources, "Oil Sands Technology Roadmap," January 2004.

16. Nick Snow, "BLM Awards Two Colorado RD&D Oil Shale Projects," *Oil & Gas Journal,* September 5, 2012; "ExxonMobil Exploration Company ("ExxonMobil") RD&D Lease Plan of Operations," and "Natural Soda Plan of Operations, Oil Shale Research, Development & Demonstration (RD&D), Tract COC 74299," both of which may be found at: http://www.blm.gov/co/st /en/fo/wrfo/Oil_Shale_-_Round_2.html (accessed February 22, 2013).

CHAPTER 2

1. Nucor presentation at the Goldman Sachs Annual Steel Conference, November 28, 2012; additional background from annual reports and other official filings.

2. IHS, "America's New Energy Future: The Unconventional Oils and Gas Revolution and the US Economy, Volume I, National Economic Contributions," *An IHS Report,* October 2012.

3. Jane Nakano et al., "Prospects for Shale Gas Development in Asia: Examining Potentials and Challenges in China and India," Center for Strategic and International Studies (CSIS), August 2012; Svetlana Izrailova, "Shale Gas Development in China," *e-International Relations,* January 9, 2013.

4. American Chemistry Council, "Shale Gas and New Petrochemical Investment: Benefits for the Economy, Jobs, and US Manufacturing," March 2011.

5. American Chemistry Council, "Shale Gas, Competitiveness and New U.S. Investment: A Case Study of Eight Manufacturing Industries," May 2012.

6. Citi GPS, "Energy 2020: North America, the New Middle East?" March 20, 2012.

7. "Comments of the Dow Chemical Company Before the United States Department of Energy, Office of Fossil Energy," 2012 LNG Export Study, January 24, 2013.

8. PWC, "Shale Gas: a Renaissance in US Manufacturing?" (December 2011); Steve James, "Analysis: Steelmakers Eye Gas to Cut Costs, Drive Exports," *Reuters,* March 16, 2012; Liam Denning, "Kafkaesque Twist in U.S.-Europe Energy Gap," *Wall Street Journal,* December 25, 2012; John M. Broder and Clifford Krauss, "A Big, and Risky, Energy Bet," *New York Times,* December 17, 2012.

9. Harold L. Sirkin, Michael Zinser, and Douglas Hohner, "Made in America, Again: Why Manufacturing Will Return to the U.S.," Boston Consulting Group, August 25, 2011; Harold L. Sirkin, Michael Zinser, Douglas Hohner, and Justin Rose, "U.S. Manufacturing Nears the Tipping Point: Which

Industries, Why, and How Much?" Boston Consulting Group, March 22, 2012; Harold L. Sirkin et al., "Why America's Export Surge Is Just Beginning," Boston Consulting Group, September 21, 2012, https://www.bcgperspectives.com /search/?SearchQuery=made%20in%20america.

10. Justin R. Pierce and Peter K. Schott, "The Surprisingly Swift Decline of U.S. Manufacturing Employment," *NBER Working Paper 18655* (December 2012). This paper finds a strong association with the granting of permanent normal trade relations with China in 2000, and hypothesizes that it eliminated uncertainty among Chinese officials about the US reaction.

11. Department of Energy (DOE), "Natural Gas Regulatory Program," see http://fossil.energy.gov/programs/gasregulation/ (accessed February 22, 2013).

12. Peter Kelly-Detwiller, "LNG Export Economics: A Look at Frontrunner Cheniere," *Forbes,* December 5, 2012, http://www.forbes.com/sites/peter detwiler/2012/12/05/lng-export-economics-a-look-at-frontrunner-cheniere /; Jane Wardell, "Exxon's PNG LNG Project Costs Balloon to $19 Billion, *Reuters,* November 11, 2012, http://www.reuters.com/article/2012/11/12/us -exxon-png-idUSBRE8AA0GR20121112.

13. Kenneth B. Medlock III, "U.S. LNG Exports: Truth and Consequence," James A. Baker III Institute for Public Policy, Rice University (August 10, 2012), 15.

14. NERA Economic Consulting, "Macroeconomic Impacts of LNG Exports from the United States," submitted to the Office of Fossil Energy, US Department of Energy, December 3, 2012, see http://www.fossil.energy.gov/ programs/gasregulation/reports/nera_lng_report.pdf.

15. Michael Levi, "A Strategy for U.S. Natural Gas Exports," The Hamilton Project, Discussion Paper 2012–04 (June 2012). This is an extremely fair and rounded presentation.

16. NERA, 62, 55.

17. "Comments of the Dow Chemical Company, Before the United States Department of Energy Office of Fossil Energy," 2012 LNG Export Study, January 24, 2013, http://www.sutherland.com/files/upload/DowChemical-Comments.pdf.

18. National Institute of Economic and Industry Research, "Large Scale Export of East Coast Australia Natural Gas: Unintended Consequences" (October 2012), ii. See "Australia Natural Gas: Unintended Consequences," available at: http://www.americasenergyadvantage.org.

19. Zack Coleman, "More Than 100 in House Press Administration to Allow Gas Exports," *The Hill E2 Wire,* January 25, 2013, http://the-hill.com/blogs/e2-wire/e2-wire/279309-lawmakers-press-energy-depart-ment-to-expand-lng-exports; Nick Snow, "US DOE Moves Carefully on

LNG Export Requests, NARUC Meeting Told," and "More Than Gas Is Needed to Control GHGs, NARUC Committee Told," *Oil & Gas Journal* (February 11, 2013): 12, 20.

CHAPTER 3

1. International Energy Agency (IEA), *Golden Rules for a Golden Age of Gas: World Energy Outlook Special Report on Unconventional Gas* (Paris, OECD /IEA, 2012), http://www.iea.org/media/WEO_GoldenRules_ForA_Golden AgeOfGas_Flyer.pdf.

2. Christopher Helman, "The Two Sides of Aubrey McClendon, America's Most Reckless Billionaire," *Forbes,* October 5, 2011; Agustino Fontevecchia, "Ex-Billionaire McClendon Out at Chesapeake Over Differences," *Forbes,* January 29, 2013.

3. Nicholas Kusnetz, "North Dakota's Oil Boom Brings Damage Along With Prosperity," *ProPublica,* June 7, 2012.

4. Alan Krupnik, Hal Gordon, and Sheila Olmstead, "Pathways to Dialogue: What the Experts Say About the Environmental Risks of Shale Gas Development," Resources for the Future, February 2013.

5. Glenn D. Schiable and Marcel P. Aillery, "Water Conservation in Irrigated Agriculture: Trends and Challenges in the Face of Emerging Demands," *Economic Information Bulletin Number 99,* Economic Research Service, US Department of Agriculture (September 2012), 28; Jack Healey, "For Farms in the West, Oil Wells Are Thirsty Rivals, *New York Times,* September 5, 2012.

6. Erick Mielke, Laura Diaz Anadon, and Venkatesh Narayanamurti, "Water Consumption of Energy Resource Extraction, Processing and Conversion: A Review of the Literature," *Discussion Paper No. 2010–15, Energy Policy Innovation Discussion Paper Series,* Harvard Kennedy School (October 2010), 6, 13–22. Lean shale gas also requires the least processing of any of the hydocarbon fuels.

7. Jack Healey, "With Ban on Drilling Practice, Town Lands in Thick of Dispute," *New York Times,* November 25, 2012.

8. Resources for the Future, Center for Energy Economics and Policy, "A Review of Shale Gas Regulations by State: Setback Restriction from Buildings; Setback Restrictions from Water Sources," available online only, http://www.rff.org/centers/energy_economics_and_policy/Pages/Shale_Maps.aspx (accessed February 22, 2013).

9. Geoff Liesik, "Utah Natural Gas Well Fire Forces Evacuations," *Deseret Morning News,* January 23, 2013.

10. Charles G. Groat and Thomas W. Grimshaw, "Fact-Based Regulation for Environmental Protection in Shale Gas Development," Energy Institute of the University of Texas at Austin (February 2012), 26.

11. Railroad Commission of Texas, "All Blowouts and Well Control Problems, All Blowouts 2011–2015," http://www.rrc.state.tx.us/data/drilling/blowouts/allblowouts11–15.php *(accessed February 22, 2013)*. 2012 data was not yet posted.

12. Stephen G. Osborn et al., "Methane Contamination of Drinking Water Accompanying Gas-Well Drilling and Hydraulic Fracturing," *PNAS* 108:20 (May 17, 2011): 8172–8176.

13. US Environmental Protection Agency (EPA), Office of Research and Development, "Investigation of Ground Water Contamination near Pavillion, Wyoming," draft paper, December 2012: xii-xiii, 1–4, 26, 29–30. The EPA maintains a web site with a rich set of documents and comments by interested parties at http://www.epa.gov/region8/superfund/wy/pavillion/. Even Terry Engelder, a professor of geoscience at Pennsylvania State University, who has played a major role in the evolving technology of shale exploitation and is assumed to be a friend of the industry, has criticized them for refusing to concede fault in obvious situations. Terry Engelder, "A Gusher of Hogwash," *Philadelphia Inquirer,* April 28, 2012.

14. Secretary of Energy Advisory Board, Shale Gas Production Subcommittee, "Ninety-Day Report," August 18, 2011, 24.

15. Resources for the Future, "A Review of Shale Gas Regulations," op. cit.; USEPA, "Methane to Markets: Reducing Methane Emissions in Pipeline Maintenance and Repair," Technology Transfer Workshop (November 2008); and "Installing Vapor Recovery Units," Technology Transfer Workshop (August 2009). See collection at http://epa.gov/gasstar/documents/workshops/. Gabrielle Pétron et al., "Estimation of Emissions from Oil and Natural Gas Operations in Northeastern Colorado," Earth System Research Laboratory, National Oceanic & Atmospheric Administration, http://www.epa.gov/ttnchie1/conference/ei20/session6/gpetron.pdf. The document is undated, but a version was submitted to the peer-reviewed *Journal of Geophysical Letters* in October 2011 and published in February 2012.

16. IEA, *Golden Rules,* 2012.

17. Robert W. Howarth, Renee Santoro, and Anthony Ingraffea, "Methane and the Greenhouse-Gas Footprint of Natural Gas from Shale Formations," *Climatic Change* 106 (2011): 679–690.

18. Timothy J. Skone, "Life Cycle Greenhouse Gas Analysis of Natural Gas Extraction & Delivery in the United States," Office of Strategic Energy Analysis and Planning, National Energy Technology Laboratory, May 12, 2011. According to Ingraffea, he and Howarth had run an energy-related lecture series and presented their paper at their concluding session. A short time later they were surprised that another lecture had been scheduled as part of the same

series. It had been organized by a colleague in order to introduce the federal refutation. The federal people, he says, told him that they were instructed to organize it and had only a couple of weeks to pull it together. Their lecture was basically a compilation of the government's standing position on the issue—which the Cornell paper was purporting to question—and makes no mention of the Cornell paper. It was widely publicized in the industry press.

19. My thanks to Mark Brownstein, chief counsel for the Environmental Defense Fund's US Energy & Climate Program, and Laura Whittenberg, who pulled together material for me. An EDF-sponsored study—Ramón A. Alvarez et al., "Greater Focus Needed on Methane Leakage from Natural Gas Infrastructure," *PNAS Early Edition* (February 13, 2012): 1–6—illustrates the importance of leakages. For University of Texas at Austin and West Virginia collaborations, see http://www.engr.utexas.edu/news/7416-allenemissions-study and http://wvutoday.wvu.edu/n/2013/03/04/scemr-release.

20. Nicholas Stern, *The Global Deal: Climate Change and the Creation of a New Era of Progress and Prosperity* (New York: PublicAffairs, 2009) is a good example. Robert Bryce, *Power Hungry: The Myths of "Green" Energy and the Real Fuels of the Future* (New York: PublicAffairs, 2010) is an intelligent countercase.

21. Dieter Helm, *The Carbon Crunch: How We're Getting Climate Change Wrong—and How to Fix It* (New Haven, CT: Yale University Press, 2012), 67–72.

22. Stern, *Global Deal,* 351.

23. Mark Z. Jacobson and Mark A. DeLucchi, "A Path to Sustainable Energy by 2030," *Scientific American* (November 2009): 58–65; for a complete exposition, see Mark Z. Jacobson and Mark A. DeLucchi, "Providing All Global Energy with Wind, Water, and Solar Power, Part I: Technologies, Energy Resources, Quantities and Areas of Infrastructure and Materials," *Energy Policy* 39 (2011): 1154–1169; and "Part II: Reliability, System and Transmission Costs, and Policies," *Energy Policy* 39 (2011): 1170–1190. Quote at 1178.

24. Helm, *Carbon Crunch,* 62–67.

PART II: PROLOGUE

1. Arthur M. Schlesinger Jr., "Arthur M. Schlesinger, Sr.: *New Viewpoints in American History* Revisited," *New England Quarterly* 61:4 (December, 1988): 483–501, at 500.

CHAPTER 4

1. Congressional Budget Office, "Public Spending on Transportation and Water Infrastructure," *A CBO Study* (November 2010): Bureau of Economic Analysis, Real Domestic GDP.

2. American Society of Civil Engineers, *America's Infrastructure Report Card for 2009* (Reston, VA: American Society of Civil Engineers, 2009), 15–18.

3. Ibid., 41–44, quote at 43.

4. Ibid., 24–29; US Environmental Protection Agency (EPA), "Drinking Water Infrastructure Needs Survey and Assessment," Fourth Report to the Congress, Washington, D.C., 2007; Congressional Budget Office, "Future Investment in Drinking Water and Wastewater Infrastructure," *A CBO Study* (November 2002).

5. ASCE, *Report Card 2009*, 33–36; US EPA, "Superfund National Accomplishments, Summary Report for 2011," http://www.epa.gov/superfund/accomp/pdfs/FY%2011%20Summary%20FINAL.pdf.

6. ASCE, *Report Card 2009*, 83–87.

7. US Department of Transportation, "2010 Status of the Nation's Highways, Bridges, and Transit: Conditions and Performance Report to Congress," Washington, D.C., March 2012: Executive Summary, 1:23, 9:28, http://www.fhwa.dot.gov/policy/2010cpr/.

8. Ibid., 5,7; ASCE, *Report Card 2009*, 101.

9. Association of American Railroads, "An Overview of America's Freight Railroads," (July 2012); Brian A. Weatherford and Henry H. Willis, "Getting America Back on Track, U.S. Public Policy on Railroads," *Policy Insight* 2 (December 2008): 8.

10. ASCE, *Report Card 2009*, 91–94.

11. Federal Aviation Administration, "Budget Estimates: Fiscal Year 2013," http://www.faa.gov/about/office_org/headquarters_offices/apl/aatf/media/FAA_FY2013_Budget_Estimates.pdf (accessed March 5, 2013).

12. Federal Aviation Administration, "NextGen Implementation Plan," Washington, D.C., 2012, http://www.faa.gov/nextgen/implementation/plan/ *(accessed March 5, 2013)*; Adrian Schofield, "FAA's NextGen Program Reaches Critical Period," *Aviation Daily*, January 14, 2013, http://www.aviationweek.com/Article.aspx?id=/article-xml/avd_01_14_2013_p05–01–534008.xml (accessed March 5, 2013).

13. Cisco Systems, Inc., "Third Annual Broadband Study Shows Global Broadband Quality Improves by 24% in One Year," October 18, 2010, http://newsroom.cisco.com/dlls/2010/prod_101710.html (*accessed* March 5, 2013). Editorial, "Why Broadband Service in the U.S. Is So Awful," *Scientific American*, October 4, 2010, http://www.scientificamerican.com/article.cfm?id=competition-and-the-internet.

14. Douglas Sutherland et al., "Infrastructure Investment: Links to Growth and the Role of Public Policies," *OECD Economics Department Working Papers*,

No. 686, OECD Publishing, March 24, 2009: 14–17 and Table 1, http://www
.oecd-ilibrary.org/economics/infrastructure-investment_225678178357.

15. Daniel Alpert, Robert Hockett, and Nouriel Roubini, "The Way For-
ward: Moving from the Post–Bubble, Post–Bust Economy to Renewed Cost
and Competitiveness," New America Foundation, October 10, 2011: 16, http://
newamerica.net/publications/policy/the_way_forward.

16. Laura Tyson, "The Case for a Multi-Year Infrastructure Investment
Plan," New America Foundation, September 6, 2010, http://www.newamerica
.net/publications/policy/the_case_for_a_multi_year_infrastructure_investment
_plan *(accessed March 5, 2013)*.

17. "Building Infrastructure: A River Runs through It," *The Economist,*
March 2, 2013.

CHAPTER 5

1. Johanna Bennett, "Top Health Care Stocks for 2013," *Barron's,* January
22, 2013.

2. William J. Baumol, *The Cost Disease: Why Computers Get Cheaper and
Health Care Doesn't* (New Haven, CT: Yale University Press, 2012); Katherine
Baicker and Jonathan S. Skinner, "Health Care Spending and the Future of
U.S. Tax Rates," *NBER Working Paper16772* (February 2011).

3. Calculated by author from: "The Pharmaceutical Industry in the United
States," http://selectusa.commerce.gov/industry-snapshots/pharmaceutical
-industry-united-states (accessed March 17, 2013); "Medical Devices: Industry
Assessment," http://ita.doc.gov/td/health/medical%20device%20industry%20
assessment%20final%20ii%203–24–10.pdf; and Ross Eisenbrey, "Health Insurance Industry Employment Outpacing Providers and All-Industry Growth
Rates," Economic Policy Institute, September 18, 2007, http://www.epi.org
/publication/webfeatures_snapshots_20070919/ (accessed March 17, 2013).

4. Baumol, *The Cost Disease,* 2012.

5. National Center for Health Statistics, "Health, United States, 2010:
With Special Feature on Death and Dying," Hyattsville, MD, February 2011,
Table 24: 137. Calculations by the author.

6. David M. Cutler et al.,"Are Medical Prices Declining: Evidence from
Heart Attack Treatments," *Quarterly Journal of Economics* 113:4 (November,
1998), 992–1023.

7. The McKinsey Global Institute,"Accounting for the Cost of US Health
Care: A New Look at Why Americans Spend More," McKinsey & Co., December 2008.

8. Antonio P. Legoretto et al., "Increased Cholecystectomy Rate after the Introduction of Laparoscopic Cholecystectomy," *Journal of the American Medical Association* 270:12 (September 22/29, 1993): 1429–1432.

9. Richard G. Frank, "Behavioral Economics and Health Economics," *NBER Working Paper 10881* (October 2004). This is an important paper. My observations of the cardiac surgery billing system at Columbia University Medical Center bear out Frank's observations. The unit expressly did not engage in profit-maximizing behavior; their approach instead fit better with a game-theory "minimax" model. They evolved a billing practice that was designed to allow them to practice their desired mode of surgery with a minimum economic penalty. As a practical matter, each year's billing strategy was worked out by the division director and the billing staff; the individual surgeons were mostly oblivious of the details.

10. American Council on Science and Health, "Wide Variations in Rates of C-Section—But Why?," March 5, 2013, http://www.acsh.org/wide-variation-in-rates-of-c-section-but-why/ (accessed March 15, 2013).

11. Atul Gawande, "The Checklist," *The New Yorker,* December 10, 2007; Jonathan Skinner and Douglas Staiger, "Technology Diffusion and Productivity Growth in Health Care," *NBER Working Paper 14865* (April 2009). It is useful to read this paper in conjunction with the Frank paper cited above.

12. Atul Gawande, "The Cost Conundrum: What a Texas Town Can Teach Us about Health Care," *New Yorker,* June 1, 2009.

13. Ibid.

14. Lawrence B. McCullough et al., eds., *Surgical Ethics* (New York: Oxford University Press, 1998), 351.

15. Reed Abelson, "Financial Ties Are Cited as Issue in Spine Study," *New York Times,* January 30, 2008.

16. Jean M. Mitchell, "Urologists' Self-Referral for Pathology or Biopsy Specimens Linked to Increased Use and Lower Prostate Cancer Detection," *Health Affairs* 31:4 (April 2012): 741–749.

17. Steven Brill, "Bitter Pill: Why Medical Bills Are Killing Us," *Time,* March 4, 2013.

18. Laura Beil, "Questions Linger for Best Use of Proton Beam Therapy," *Cure,* September 13, 2012, http://curetoday.com/index.cfm/fuseaction/article.show/id/2/article_id/1976.

19. Department of Justice, Office of Public Affairs, "Justice Department Recovers Nearly $5 Billion in False Claims Act Cases in Fiscal Year 2012," December 4, 2012, http://www.justice.gov/opa/pr/2012/December/12-ag-1439.

html; Skadden, "Pharma/Device Enforcement: A Year in Review," January 19, 2012.

20. Frank R. Lichtenberg, "The Impact of Therapeutic Procedure Innovation on Hospital Patient Longevity: Evidence from Western Australia, 2000–2007," *NBER Working Paper 17414* (September, 2011). This article also includes a summary of previous work by other authors. Frank R. Lichtenberg, "Have Newer Cardiovascular Drugs Reduced Hospitalization? Evidence from Longitudinal Country-Level Data on 20 OECD Countries, 1995–2003," *NBER Working Paper 14008* (May 2008); Lichtenberg, "The Quality of Medical Care, Behavioral Risk Factors, and Longevity Growth," *NBER Working Paper 15068* (June 2009); Lichtenberg, "Has Medical Innovation Reduced Cancer Mortality?" *NBER Working Paper 15880* (April 2010); Lichtenberg, "The Effect of Pharmaceutical Innovation on the Functional Limitations of Elderly Americans: Evidence from the 2004 Nursing Home Survey," *NBER Working Paper 17750* (January 2012); Lichtenberg, "Pharmaceutical Innovation and Longevity Growth in 30 Developing and High-Income Countries, 2000–2009), *NBER Working Paper 18325* (July 2012).

21. See Charles Morris, "The Measurement Problem," in *The Surgeons: Life and Death in a Top Heart Center* (New York: W. W. Norton & Co., 2007), for an extended treatment.

22. OECD, "Health: Key Tables from OECD—Total Expenditure on Health Per Capita at Current Prices and PPPs," October 30, 2012, http://www.oecd-ilibrary.org/social-issues-migration-health/total-expenditure-on-health-per-capita-2012-2_hlthxp-cap-table-2012-2-en (accessed March 17, 2013).

CHAPTER 6

1. Congressional Budget Office, "The Budget and Economic Outlook: Fiscal Years 2013 to 2023," February 2013.

2. Organization for Economic Co-operation and Development (OECD), "Economic Output, Analysis and Forecasts: Long-Term Baseline Projections," http://stats.oecd.org/BrandedView.aspx?oecd_bv_id=eo-data-en&doi=data-00645-en (accessed March 15, 2013).

3. Uwe E. Reinhardt, Peter S. Hussey, and Gerard F. Anderson, "U.S. Health Care Spending in an International Context," *Health Affairs* 23:3 (May-June 2004): 10–25, at 19–20.

4. Mari Iwata, "Chevron: Most LNG Prices to Remain Linked to Oil," *Wall Street Journal,* December 5, 2012.

5. Charles River Associates, "US Manufacturing and LNG Exports: Economic Contributions to the US Economy and Impacts on US Natural Gas

Prices," February 25, 2013, 22, 33–34. The report was financed by the Dow Chemical Company. It is available online at http://www.lnglawblog.com /03152013dow/.

6. Stockholm International Peace Research Institute, "SIPRI Military Expenditure Data Base" (updated through 2011), http://www.sipri.org/databases /milex (accessed March 15, 2013).

7. Ibid. Calculation by author.

8. OECD, "Revenue Statistics—Comparative Tables," http://stats.oecd.org /Index.aspx?QueryId=21699 (accessed March 15, 2013).

Index

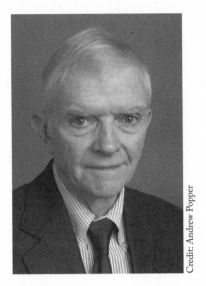

CHARLES R. MORRIS has written many books, including *The Cost of Good Intentions*, a *New York Times* Best Book of 1980; *The Coming Global Boom*, a *New York Times* Notable Book of 1990; *The Tycoons*, a *Barrons'* Best Book of 2005; and *The Trillion-Dollar Meltdown*, winner of the Gerald Loeb Award and a *New York Times* Bestseller. A lawyer and former banker, Mr. Morris's articles and reviews have appeared in many publications including the *Atlantic Monthly*, the *New York Times*, and the *Wall Street Journal*. He is a fellow of the Century Foundation and a member of the Council on Foreign Relations.

PublicAffairs is a publishing house founded in 1997. It is a tribute to the standards, values, and flair of three persons who have served as mentors to countless reporters, writers, editors, and book people of all kinds, including me.

I. F. STONE, proprietor of *I. F. Stone's Weekly*, combined a commitment to the First Amendment with entrepreneurial zeal and reporting skill and became one of the great independent journalists in American history. At the age of eighty, Izzy published *The Trial of Socrates*, which was a national bestseller. He wrote the book after he taught himself ancient Greek.

BENJAMIN C. BRADLEE was for nearly thirty years the charismatic editorial leader of *The Washington Post*. It was Ben who gave the *Post* the range and courage to pursue such historic issues as Watergate. He supported his reporters with a tenacity that made them fearless and it is no accident that so many became authors of influential, best-selling books.

ROBERT L. BERNSTEIN, the chief executive of Random House for more than a quarter century, guided one of the nation's premier publishing houses. Bob was personally responsible for many books of political dissent and argument that challenged tyranny around the globe. He is also the founder and longtime chair of Human Rights Watch, one of the most respected human rights organizations in the world.

. . .

For fifty years, the banner of Public Affairs Press was carried by its owner Morris B. Schnapper, who published Gandhi, Nasser, Toynbee, Truman, and about 1,500 other authors. In 1983, Schnapper was described by *The Washington Post* as "a redoubtable gadfly." His legacy will endure in the books to come.

Peter Osnos, *Founder and Editor-at-Large*